Nelson Advanced Science

Fields, Forces and Synthesis

Mark Ellse • Chris Honeywill

Published in 2001 by:
Nelson Thornes Ltd
Delta Place
27 Bath Road
CHELTENHAM
GL53 7TH
United Kingdom

01 02 03 04 05 / 10 9 8 7 6 5 4 3 2

A catalogue record for this book is available from the British Library

ISBN 0 17 448300 7

Picture research by Zooid
Illustrations and page make-up by Hardlines

Printed and bound in Croatia by Zrinski d.d. Cakovec

Acknowledgements

The authors and publisher are grateful for permission to include the following copyright material:

The examination questions are reproduced by permission of London Examinations, a division of Edexcel Foundation.

Photographs:
C.T.R. Wilson/SPL: figure 23.3
Corbis UK Ltd: p.1, Dennis di Cicco; p.47, Kevin Fleming
David Parker/University of Birmingham High TC Consortium/SPL: p.33
djb microtech: figure 11.2
Hulton Getty Picture Collection Ltd: figure 21.3
John Mason/SPL: figure 24.3
Martyn F. Chillmaid: p.19
Peter Gould: figures 1.1, 1.2, 2.3, 5.1, 5.3, 13.4(a), 15.1, 15.4, 17.1, 22.2
Science Photo Library: figures 3.2, 21.4, 23.1, 23.2

Contents

Introduction

This series has been written by Principal Examiners and others involved directly with the development of the Edexcel Advanced Subsidiary (AS) and Advanced (A) GCE Physics specifications.

Fields, Forces and Synthesis is one of four books in the Nelson Advanced Science (NAS) series developed by updating and reorganising the material from the Nelson Advanced Modular Science (AMS) books to align with the requirements of the Edexcel specifications from September 2000. The books will also be useful for other AS and Advanced courses.

Fields, Forces and Synthesis covers both Unit 5 of the A2 physics specification and Unit 6. Unit 5 is concerned with the study of Fields. Unit 6 is concerned with Synthesis – the bringing together of all the different parts of the physics specification.

Many topics on the synthesis paper draw heavily on the concepts of fields in Unit 5. we have therefore chosen to cover these synthesis topics alongside the related field topics of Unit 5. We have marked the pages that relate to the synthesis topic with a BLUE SYNTHESIS WATERMARK across the relevant chapters. The last topic in this book – accelerators – relates to synthesis material alone.

Other resources in this series

NAS Teachers' Guide for AS and A Physics provides a proposed teaching scheme together with practical support and answers to all the practice and assessment questions provided in *Mechanics and Radioactivity; Electricity and Thermal Physics; Waves and Our Universe;* and *Fields, Forces and Synthesis*.

NAS Physics Experiment Sheets 2nd edition by Adrian Watt provides a bank of practical experiments that align with the NAS Physics series. They give step-by-step instructions for each practical provided and include notes for teachers and technicians.

NAS Make the Grade in AS and A Physics is a Revision Guide for students. It has been written to be used in conjunction with the other books in this series. It helps students to develop strategies for learning and revision, to check their knowledge and understanding and to practise the skills required for tackling assessment questions.

Features used in this book

The Nelson Advanced Science series contains particular features to help you understand and learn the information provided in the books, and to help you to apply the information to your studies.

These are the features that you will find in the Nelson Advanced Science Physics series. (All examplar extracts are taken from the first book in the series.)

Text offers complete coverage of content. It is divided into small, self-contained chapters, each of which introduces only a few new ideas. Important terms are indicated in **bold**.

15 Types of forces

Figure 15.1   There is a gravitational force between the spacecraft and the Earth

Forces acting at a distance

The forces that bodies in space are most likely to experience are gravitational. Imagine a spacecraft near to a planet, as in Figure 15.1. The planet pulls the spacecraft towards the planet (Figure 15.2). And, just as you would expect, the spacecraft pulls the planet with an equal force towards the spacecraft (Figure 15.3).

The force on the spacecraft is equal and opposite to the force on the planet, and the forces have the same line of action and act for the same time.

The two bodies are exerting a force on each other without being in contact. These are gravitational forces; they act at a distance. Gravitational forces occur between all bodies of all sizes and attract them together. Gravitational forces are one of the four different types of force that exist.

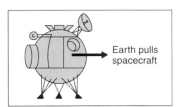

Earth pulls spacecraft

Figure 15.2   The Earth pulls the spacecraft right, so the spacecraft accelerates to the right

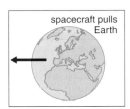

spacecraft pulls Earth

Figure 15.3   The spacecraft pulls the Earth left, so the Earth accelerates to the left

Gravity and gravitational fields

Masses around any body experience gravitational forces. The region in which they experience this force is the body's gravitational field. The gravitational force that a planet applies to a body is called its **weight**. When you are standing on the Earth, the Earth pulls you towards it and provides your weight. You pull the Earth with an equal and opposite force.

INTRODUCTION

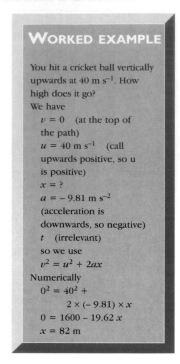

Worked examples provide step-by-step guidance on how to do the calculations required for some of the areas of physics.

Practical experiments are suggested within the text. Though you may only do some of these or see them demonstrated, you should think about them all as you read about them and try to predict what the experiments would show.

Using ticker timers and video cameras to measure acceleration

- A ticker timer puts 50 ticks on a tape per second. Let a trolley pull a tape through the ticker timer as it runs down the runway (Figure 6.3).
- Calculate the average velocity at 0.1 s intervals (Figure 6.4) and plot a graph of velocity against time to calculate the acceleration.
- A video camera takes 25 pictures per second. Video your trolley running down the runway, with a calibrated scale in the background. Play the video back a frame at a time and measure from the screen the distance travelled per frame.
- Compare these results with those using the ticker timer.

Figure 6.3   Trolley with ticker timer

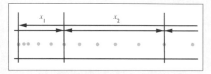

Figure 6.4   Mark the start of the tape (first tick) and tabulate values of the distance every five ticks afterwards. This is the distance covered each tenth of a second

Practice questions are provided towards the end of the book. Several of these are provided for every chapter and will give you the opportunity to check your knowledge and understanding after studying each chapter.

Typical assessment questions are found at the end of the book. These are similar in style to the assessment questions for Advanced GCE that you will encounter in your Unit Tests (exams) and they will help you to develop the skills required for these types of questions.

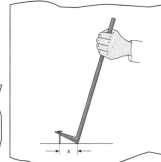

Things you need to know provides a list of terms shown in **bold** in the text, with definitions. They are organised by chapter at the end of the book.

Acknowledgements

John Warren and David Hartley painstakingly read through and commented in detail on the manuscripts of the first edition. The authors and publisher gratefully acknowledge their major contributions to the success of the whole series.

Mark Burton and Frances Kirkman gave kind assistance during the preparation of the new Edexcel specification and advised on the provision of the assessment questions.

About the authors

Mark Ellse is Principal of Chase Academy in Cannock, Staffordshire, and Principal Examiner for Edexcel.

Chris Honeywill is Assistant Principal Examiner for Edexcel and former Head of Physics at Farnborough Sixth Form College.

Fields

Electricity, gravitation and magnetism affect things that are some distance away. They have **fields** — regions in which they exert forces. In a star, the huge gravitational forces crush protons against their electrostatic repulsion, to form higher elements from single hydrogen nuclei.

Huge gravitational forces are responsible for the fusion process that occurs in all stars, including those in the Andromeda galaxy.

Gravitational fields

Figure 1.1 A 55 kg woman and a 70 kg man

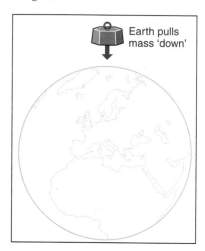

Figure 1.3 The mass accelerates because of the single force on it

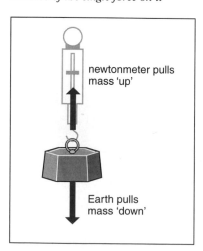

Figure 1.4 The force of the newtonmeter is equal to the weight

Mass

Perhaps the most basic quantity in physics is **mass**. The mass of a body is the amount of matter in that body. Mass is a base quantity measured in kilograms. The man and woman in Figure 1.1 are about average size; their masses are about 70 kg for the man and 55 kg for the woman. Mass is a scalar quantity – it has no direction associated with it.

The mass of a body does not vary from place to place. The masses of the two people in Figure 1.1 are always the same, whether on the Earth, on the Moon, or floating about in outer space.

Measuring weight

- Hang a 1 kg mass on a newton spring balance (Figure 1.2). Measure the force required to support the mass.
- Repeat with masses of 0.5 kg and 2 kg.
- What is the resultant force on the mass when it is hanging on the end of the balance? Draw a free-body force diagram for the mass.

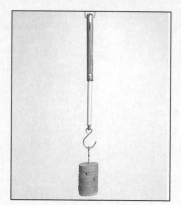

Figure 1.2 The weight of 1 kg is 9.8 N

Weight

If you release a body near the Earth, it accelerates towards the Earth, because the Earth exerts a gravitational force on the body (Figure 1.3). This gravitational force is the body's **weight**.

When you hang a body on a spring balance, the spring balance provides the upward force to support the body. The force of the balance on the body is equal and opposite to the weight, as shown in Figure 1.4. The resultant of this contact force from the balance and the weight is zero, so the body is in equilibrium. It hangs at rest from the balance.

Gravitational fields

Physicists use the word **field**, or **force field**, for a region in which forces act. Any mass near the Earth experiences a force of gravity towards the Earth. This region, where the Earth exerts a gravitational force on a mass, is the Earth's **gravitational field**. Other bodies have gravitational fields. The Sun's gravitational field is responsible for the orbits of the planets. The Moon's gravitational field affects the tides on Earth.

Gravitational field strength

In the experiment 'Measuring weight', you learnt that the Earth pulls harder on large masses than on small ones. The weight of an object depends on the mass of that object. Near the Earth's surface, the Earth pulls on each kilogram with a force of about 9.8 N, and proportionately more or less on larger or smaller masses.

The **gravitational field strength** (g) is the force exerted by a gravitational field on each kilogram. Near the Earth the gravitational field strength is about 9.8 N kg^{-1}. So

$$\text{weight} = \text{mass} \times \text{gravitational field strength} \qquad F = mg$$

For a woman of mass 55 kg near the Earth,

$$F = mg = 55 \text{ kg} \times 9.8 \text{ N kg}^{-1} = 540 \text{ N}$$

Her weight on Earth is 540 N.

The Earth's gravitational field strength depends on distance from the Earth. As you get higher, the field strength gets less. If you take an accurate spring balance and a 1 kg mass from sea level to a height of 8 km, you can measure that g has fallen from an average value of 9.81 N kg^{-1} to 9.78 N kg^{-1}. At the distance of the Moon, the gravitational field strength of the Earth has dropped to 0.0028 N kg^{-1}. You can see how to calculate this value in Chapter 4.

Field strength and acceleration

You know from *Mechanics and Radioactivity* that a body in free fall near the Earth accelerates at a rate of about 9.8 m s^{-2}. It is not a coincidence that this is the same numerical value as the gravitational field strength.

The mass in Figure 1.3 is falling freely. It has only its weight acting on it:

$$\text{weight} = \text{mass} \times \text{gravitational field strength}$$

The body's acceleration can be calculated from Newton's second law:

$$\text{force (the weight)} = \text{mass} \times \text{acceleration}$$

So equating these two expressions for weight gives

$$\text{mass} \times \text{gravitational field strength} = \text{mass} \times \text{acceleration}$$

$$\text{gravitational field strength} = \text{acceleration}$$

$$g = a$$

You can show that the units of g and a are the same. The unit of g is N kg^{-1} = (kg m s^{-2}) kg^{-1} = m s^{-2}, which is the unit of a.

Newton's law of gravitation

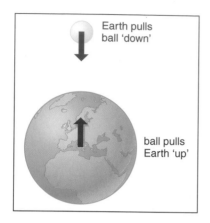

Figure 2.1 The Earth attracts the ball, and the ball attracts the Earth

Mutual attraction

It is no great surprise that a body has weight: the Earth pulls the body down. But it is not so obvious that the body also pulls the Earth up with an equal and opposite force. Figure 2.1 shows the forces between a ball and the Earth. You studied this situation in Chapters 14 and 16 of *Mechanics and Radioactivity*.

Weight, a gravitational force, is not a single force acting on one object, but one of a pair of forces that act between two objects. Gravitational forces are the result of *mutual* attraction, where two objects both attract each other.

Isaac Newton suggested theories about gravitation based on astronomical observations. He suggested that gravitational forces occur between all pairs of bodies. There are gravitational forces between you and this book, as well as between this book and the Earth (Figure 2.2).

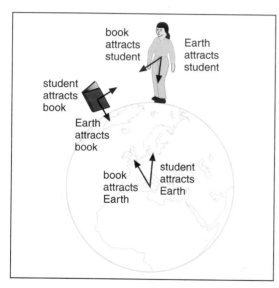

Figure 2.2 There are forces between you and the book, as well as between the Earth and the book

Investigating the forces between two masses

- Place a lead mass on top of a sensitive balance and note the reading.
- Observe the balance reading as you bring up another lead mass supported from a retort stand (Figure 2.3).
- Comment on the size of the gravitational forces between the two masses.

Figure 2.3 Any change in the balance reading shows the force between the two masses

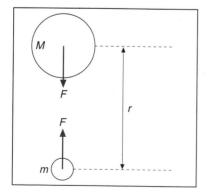

Figure 2.4 Every mass attracts every other mass

Newton's law of gravitation

Newton's law of gravitation states that every mass attracts every other mass. The force of attraction is proportional to each of the masses and inversely proportional to the square of the distance apart:

$$F = \frac{GmM}{r^2}$$

where F is the force, m and M the two masses and r the distance apart (Figure 2.4). G is a constant that applies in all situations of gravitational attraction. It is called the **universal gravitational constant**. Its value is 6.67×10^{-11} N m^2 kg^{-2}.

Newton's law of gravitation applies to all spherical masses if you measure the distance apart between their centres of mass. The law also applies to non-spherical masses if their distance apart is large compared with the size of the masses.

WORKED EXAMPLE

Estimate the gravitational force between the two people, masses 55 kg and 70 kg, in Figure 1.1 if they stand with their centres of mass 0.75 m apart.
We get

$$F = GmM/r^2$$

$$= 6.67 \times 10^{-11} \text{ N m}^2 \text{ kg}^{-2} \times 55 \text{ kg} \times 70 \text{ kg}/(0.75 \text{ m})^2$$

$$= 4.6 \times 10^{-7} \text{ N} = 0.47 \, \mu\text{N}$$

Measuring G

Newton's law of gravitation predicts that gravitational force decreases if the objects are further apart. If you are a long way from the Earth, your weight is less, because the gravitational force is less.

If one or both masses are small, the gravitational forces of attraction are small. The attractive force between a pair of masses in the laboratory is far too small to measure with an electronic top-pan balance.

Over a hundred years after Newton proposed his law, Henry Cavendish devised the first practical method of measuring the gravitational force between masses. He attached two small lead masses to a bar and hung it from a long thin wire (Figure 2.5). As well as supporting the bar and masses, the wire acted like a very weak spring, rotating the bar back to its equilibrium position. He recorded the equilibrium position of the bar. Then he placed two larger lead masses in the positions shown. They attracted the masses on the bar, and caused it to rotate slightly against the restoring force from the wire. He measured the angle through which the bar rotated, and from this calculated the force between the masses. By measuring the forces, the masses and their distances apart, he calculated G.

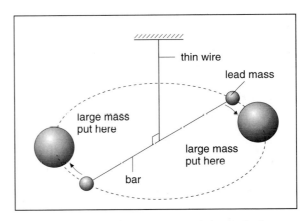

Figure 2.5 Cavendish's method

Cavendish realised that measuring G allowed him to calculate the mass of the Earth. Think about a 1 kg mass on the Earth's surface. The weight of this mass is 9.8 N and, since the radius of the Earth is 6400 km, it is this distance from the centre of the Earth.
So rewriting the equation $F = GmM/r^2$ gives

$$M = Fr^2/Gm$$

$$= 9.8 \text{ N} \times (6.4 \times 10^6 \text{ m})^2/(6.7 \times 10^{-11} \text{ N m}^2 \text{ kg}^{-2} \times 1 \text{ kg}) = 6.0 \times 10^{24} \text{ kg}$$

Satellites

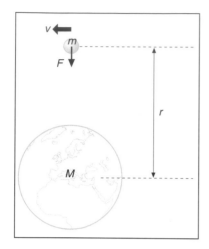

Figure 3.1 The Earth provides the centripetal force for the Moon's circular motion

Planets and satellites

You know from Chapter 1 of *Waves and Our Universe* that a body executing circular motion needs a centripetal force. For a planet or satellite orbiting a star or planet, the gravitational attraction provides the centripetal force.

The Moon orbits the Earth. Figure 3.1 shows a force diagram for the Moon. There is only one force F acting on the Moon. This is the gravitational force of the Earth (GmM/r^2), which provides the centripetal force (mv^2/r) for the Moon's circular motion. The single force on the Moon is the centripetal force and it is also the gravitational force. So the two mathematical ways of describing this force must be equal, i.e. centripetal force = gravitational force:

$$mv^2/r = GmM/r^2 \qquad \text{so} \qquad v^2/r = GM/r^2$$

The centripetal acceleration is v^2/r, and GM/r^2 is the gravitational field strength at radius r. So this last equation states that the centripetal acceleration is the gravitational field strength. You might like to compare this with the end of Chapter 1, which explained that the acceleration of free fall at a point is the gravitational field strength at that point. A satellite is in free fall. The acceleration of free fall is the gravitational field strength at that point, which accounts for the centripetal acceleration.

It is often useful to write the equations for satellite motion in terms of the angular speed ω. The centripetal force is $mr\omega^2$:

$$mr\omega^2 = GmM/r^2 \qquad \text{so} \qquad r\omega^2 = GM/r^2$$

The centripetal acceleration is $r\omega^2$, and GM/r^2 is the gravitational field strength at radius r. So again this last equation states that the centripetal acceleration is the gravitational field strength.

Figure 3.2 Nimbus, a near-Earth, polar satellite used for ozone distribution and environmental monitoring

WORKED EXAMPLE

The period T of the Moon about the Earth is about 28 days. Calculate the radius of the Moon's orbit around the Earth, given that the mass of the Earth is 6.0×10^{24} kg and $G = 6.7 \times 10^{-11}$ N m^2 kg^{-2}.

For circular motion questions, it is often useful to find the angular velocity ω:
$$\omega = 2\pi/T = 2\pi \text{ rad}/(28 \times 24 \times 3600 \text{ s}) = 2.6 \times 10^{-6} \text{ rad s}^{-1}$$

Now we can use centripetal acceleration = gravitational field strength
$$r\omega^2 = GM/r^2$$

$$r^3 = GM/\omega^2 = 6.7 \times 10^{-11} \text{ N m}^2 \text{ kg}^{-2} \times 6.0 \times 10^{24} \text{ kg}/(2.6 \times 10^{-6} \text{ rad s}^{-1})^2$$

$$r = \sqrt[3]{(5.95 \times 10^{25} \text{ m}^3)} = 390 \text{ Mm}$$

Kepler's law

You know that centripetal acceleration = gravitational field strength:

$$r\omega^2 = GM/r^2 \qquad \text{so} \qquad r^3\omega^2 = GM$$

But $\omega = 2\pi/T$ where T is the period of the circular motion. This gives

$$r^3(2\pi/T)^2 = GM \qquad \text{so} \qquad T^2 = (4\pi^2/GM)r^3$$

This law says that the period T of an object in gravitational circular motion depends on the mass M of the object about which it is rotating and the radius of the orbit. For the planets orbiting around the Sun (constant M), the squares of their periods are proportional to their radii cubed:

$$T^2 \propto r^3$$

This is one of a series of laws that the German astronomer Johannes Kepler discovered by analysing observations of the planets' motion. Newton based his theory of gravitation on Kepler's laws. Table 3.1 shows the periods and orbital radii of the planets.

Table 3.1 *The periods and orbital radii of the planets*

Planet	Period /Earth years	Orbital radius /Earth's orbital radius
Mercury	0.24	0.39
Venus	0.62	0.73
Earth	1.0	1.0
Mars	1.9	1.5
Jupiter	12	5.2
Saturn	29	10
Uranus	84	19
Neptune	160	30
Pluto	250	39

Satellites

An artificial satellite is an object placed into orbit around a planet, usually around the Earth. Satellites were first put into space for research. Now they are used for many purposes. Near-Earth satellites (Figure 3.2) orbit very quickly, just above the atmosphere, where g is almost the same as its value on the Earth's surface. They are used for weather observation, for mapping the ground and for spying. The period of a near-Earth satellite is about 90 min.

Geostationary (geosynchronous) satellites have a period of 24 h. They have the same period as the Earth. They orbit above the equator and keep the same position above the Earth's surface. The most important use for geostationary satellites is for telecommunications. Geostationary satellites broadcast satellite television and relay telephone conversations over wide areas of the Earth.

The radius of the orbit of a geostationary satellite is about 42×10^6 m, just over a tenth of the way from the Earth to the Moon.

Figure 3.3 shows the relative positions of the Moon and the two types of satellites above.

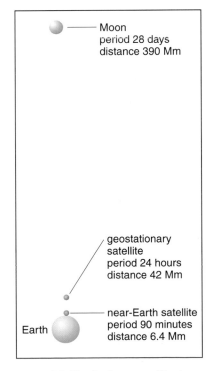

Figure 3.3 The further a satellite is from the Earth, the greater its period

4 Gravitational field lines

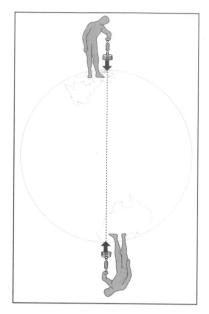

Figure 4.1 The Earth pulls masses towards its centre

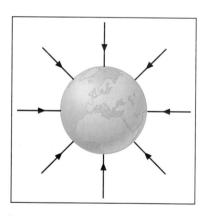

Figure 4.2 The Earth's field is radial

The shape of the Earth's field

Wherever a mass is, in England or Australia, near the ground or in space, its weight acts towards the centre of the Earth, as Figure 4.1 shows. The lines on Figure 4.2 are field lines. They show the directions of the gravitational forces at any point. The field lines are radial. They are all directed towards the centre of the Earth. Near the Earth, where the lines are close together, the field is stronger. A long way from the Earth, where the field lines are further apart, the field is weaker.

Calculating the strength of the Earth's field

If a body of mass m is a distance r from the centre of the Earth, you know from previous chapters that the weight of the body, F, is given by the formula $F = GmM/r^2$. The gravitational field strength is

$$g = F/m = (GmM/r^2)/m = GM/r^2$$

This formula allows you to calculate g at any distance from the Earth.

On the surface, 6.4×10^6 m from the centre of the Earth, $g = 9.8$ N kg^{-1}. Twice as far from the centre of the Earth (12.8×10^6 m from the centre),

$$g = GM/r^2$$

$$= 6.7 \times 10^{-11} \text{ N m}^2 \text{ kg}^{-2} \times 6.0 \times 10^{24} \text{ kg}/(12.8 \times 10^6 \text{ m})^2$$

$$= 2.5 \text{ N kg}^{-1}$$

This is a quarter of the value at the Earth's surface.

Be careful to notice the difference between G and g. G does not vary. It is the universal gravitational constant, the constant of proportionality in Newton's law of gravitation. The gravitational field strength g is the gravitational field strength at a particular point. It is the force per unit mass. It varies from place to place.

WORKED EXAMPLE

Calculate the gravitational field strength due to the Sun at the orbit of the Earth. The mass of the Sun is 2.0×10^{30} kg. The radius of the Earth's orbit is 1.5×10^{11} m.
We get

$$g = GM/r^2$$

$$= 6.7 \times 10^{-11} \text{ N m}^2 \text{ kg}^{-2} \times 2.0 \times 10^{30} \text{ kg}/(1.5 \times 10^{11} \text{ m})^2$$

$$= 6.0 \times 10^{-3} \text{ N kg}^{-1}$$

Inverse square law

You read about the **inverse square law** in Chapter 10 of *Waves and Our Universe*. Like gravitational forces, gravitational fields follow an inverse square law; $g \propto 1/r^2$. If you double the distance from the centre of a mass, the field strength quarters. If you treble the distance, the field strength reduces to a ninth. If the distance is ten times as large, the field is a hundredth.

Figure 4.3 shows a graph of the Earth's gravitational field and how it varies with distance from the centre of the Earth.

At the Moon's orbit, a distance of 380 Mm from the Earth, the gravitational field strength is only 2.8 mN kg^{-1}, over 3000 times smaller than at the surface of the Earth. But this tiny gravitational field strength is still large enough to cause the Moon to circle round the Earth, rather than continuing in a straight line through space.

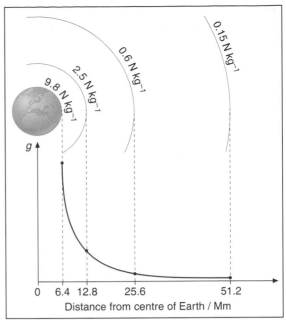

Figure 4.3 How g varies with distance from the centre of the Earth

A uniform gravitational field

In a relatively small region, for instance across a soccer pitch, across a small country, even up a high mountain, the Earth's gravitational field lines are almost parallel.

Figure 4.4b shows the small shaded section of the Earth's field from Figure 4.4a. The distance apart of the field lines is effectively constant, and the gravitational field strength also varies little in this comparatively small region. So you can consider the Earth's gravitational field within a small region to be uniform. This simplifies calculations.

Work done in a uniform gravitational field

The weight of a kilogram is 9.8 N on the Earth's surface. If you raise it 1.0 m,

> **work = force × distance = 9.8 N × 1.0 m = 9.8 J**

If you raise it 1.0 m its potential energy (PE) increases by 9.8 J. (Strictly, the mass itself does not have PE. The PE is a property of both the mass and the Earth, which pull each other together.)

To raise a mass of 1 kg through 2 m needs 9.8 N × 2 m = 19.6 J, and its PE increases by 19.6 J during this process.

For a mass m, in a gravitational field g, the force needed to raise it is mg. If you raise it through a height Δh, then

> **increase in PE = work done = force × distance = $mg \times \Delta h = mg\Delta h$**

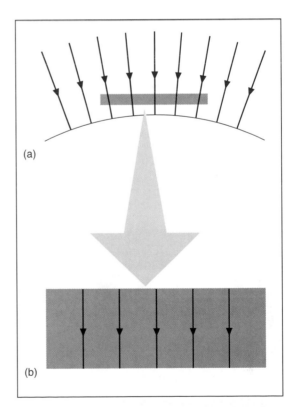

Figure 4.4 Across a relatively small region, the Earth's field is effectively uniform

5 Coulomb's law

Measuring charge

- Rub a polythene strip with a duster to charge it negatively. Then scrape some of that charge onto the top cap of a coulombmeter to measure the charge (Figure 5.1).
- Repeat the experiment using an acetate rod rubbed to charge it positively.
- Connect together the terminals of a charged coulombmeter with a piece of wire; note what happens.
- Repeat with a piece of paper. Then try pieces of a range of materials.

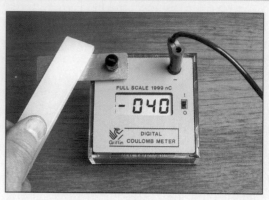

Figure 5.1 Charging a coulombmeter

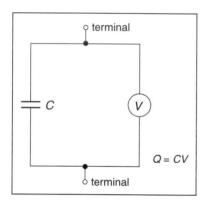

Figure 5.2 The charge $Q = CV$

The coulombmeter

You learnt in Chapter 2 of *Electricity and Thermal Physics* about two types of charge: positive and negative. You can measure charge with a coulombmeter.

Figure 5.2 shows the circuit of a coulombmeter. It is a voltmeter connected across a known capacitance. The meter is calibrated to indicate the charge Q: the capacitance C multiplied by the voltage V.

Electrons will move through a conductor placed between the terminals of a coulombmeter and discharge the capacitor. But electrons cannot move through an insulator. So if an insulator is placed between the terminals of a coulombmeter, the coulombmeter will not be discharged.

Electrons and protons

You learnt in Chapter 18 of *Waves and Our Universe* that a negatively charged object has more electrons than protons. A positively charged object has more protons than electrons. The size of the charge on a proton is the same as the size of the charge on an electron. This is called the **electronic charge**. It is negative on an electron (-1.6×10^{-19} C) and positive on a proton ($+1.6 \times 10^{-19}$ C). An object can only have a whole number of protons or electrons. So its charge must be a multiple of 1.6×10^{-19} C.

Figure 5.3 Measure the force between the two polythene rods

Forces between charges

- Place a charged polythene rod on an insulator resting on the pan of a sensitive balance and hold another charged rod near it (Figure 5.3).
- How does the direction of the force depend on the sign of the charge?
- Investigate how the size of the force depends on the distance between the charges.

Coulomb's law

The force between two charges obeys a law similar to that between two masses. The force is larger if the charges are large and larger if the charges are close together. For the two point charges q and Q, separated by a distance r in Figure 5.4, the force between them is:

$$F = \frac{kqQ}{r^2}$$

where k has the value 9.0×10^9 N m^2 C^{-2} in a vacuum or air. This formula is known as **Coulomb's law**.

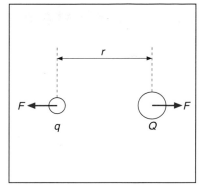

Figure 5.4 There are forces between pairs of charges

Positive and negative forces

All gravitational forces are attractive, but electrostatics forces can be either attractive or repulsive (Figure 5.5).

If both charges are positive, the formula for Coulomb's law $F = kqQ/r^2$ gives a positive value, and the force is repulsive. With two negative charges, the formula again gives a positive value, and the force is again repulsive.

With one positive charge and one negative charge, the formula gives a negative value, showing that the force is attractive.

You can use Coulomb's law to calculate the force between two spherical charges. If you charge a small conducting ball with a 5000 V power supply, it acquires a charge of about 12 nC. If two of these spheres are placed with their centres 6.0 cm apart, the force between them is:

$$F = \frac{kqQ}{r^2}$$

$$= 9.0 \times 10^9 \text{ N m}^2 \text{ C}^{-2} \times 12 \times 10^{-9} \text{ C} \times 12 \times 10^{-9} \text{ C}/(0.06 \text{ m})^2$$

$$= 0.36 \text{ mN}$$

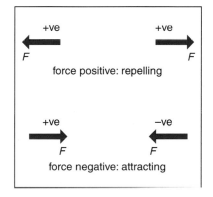

Figure 5.5 Like charges repel: the force is positive. Unlike charges attract: the force is negative

This force is small, about a hundredth of the weight of a page of this book. But the force is measurable. Electrostatic forces between small bodies are noticeable, unlike gravitational forces between small bodies.

The constant k depends on the material between the charges. It has the value 9.0×10^9 N m^2 C^{-2} in vacuum or air. Its value is calculated from the formula

$$k = \frac{1}{4\pi\varepsilon_0}$$

where ε_0 is the permittivity of a vacuum; ε_0 has the value 8.85×10^{-12} F m^{-1}.

6 | Radial electric fields

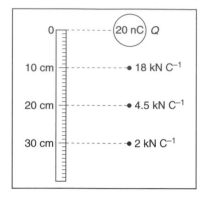

Figure 6.1 The field strength decreases as you get further from charge Q

Electric field strength

An **electric field** is a region in which there are forces on charges. As you would expect, the fields themselves are caused by charges. The **electric field strength** at a point is the force exerted by an electric field on one coulomb:

$$\text{electric field strength} = \text{force/charge} \qquad \text{or} \qquad E = \frac{F}{q}$$

The unit of electric field strength is the newton per coulomb (N C^{-1}).

If a charge q is a distance r from another charge Q, you know from the last chapter that the force F between them is given by the formula

$$F = \frac{kqQ}{r^2}$$

The charge Q has an electric field strength

$$E = \frac{F}{q} = \frac{kqQ}{r^2} / q = \frac{kQ}{r^2}$$

This formula allows you to calculate the field strength at any distance from a charge.

At a distance of 0.1 m from the centre of the charge of 20 nC in Figure 6.1,

$$E = \frac{kqQ}{r^2} = \frac{9.0 \times 10^9 \,\text{N}\,\text{m}^2\,\text{C}^{-2} \times 20 \times 10^{-9}\,\text{C}}{(0.1\,\text{m})^2} = 18\,\text{kN C}^{-1}$$

Twice as far away from the centre of the charge, the field strength is 4.5 kN C^{-1}. The electric field strength of a point charge obeys the inverse square law: if you double the distance from the charge, you quarter the field strength. Figure 6.2 is a graph of field strength against distance from Q.

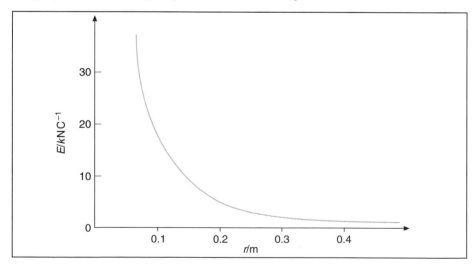

Figure 6.2 Field strength obeys an inverse square law with distance

Finding the shape of electric fields

• Connect a 5000 V power supply to two circular electrodes immersed in castor oil, as shown in Figure 6.3. Then sprinkle semolina grains on to the castor oil. Draw the pattern that results.

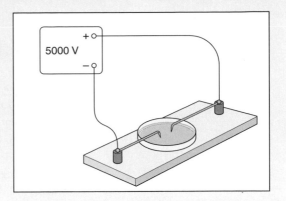

Figure 6.3 The power supply sets up an electric field between these electrodes

The field around a point charge

If you put a positive charge q anywhere near another positive charge Q, the force on q is always directed away from Q. The lines on Figure 6.4 are field lines. They show the directions of the electric force at any point due to charge Q. The field lines are radial. They are all directed out from the centre of charge.

Near Q, where the lines are close together, the field is stronger. A long way from Q, where the field lines are further apart, the field is weaker.

If Q is negative, the field lines are directed towards the charge. The field is radial, but inwards.

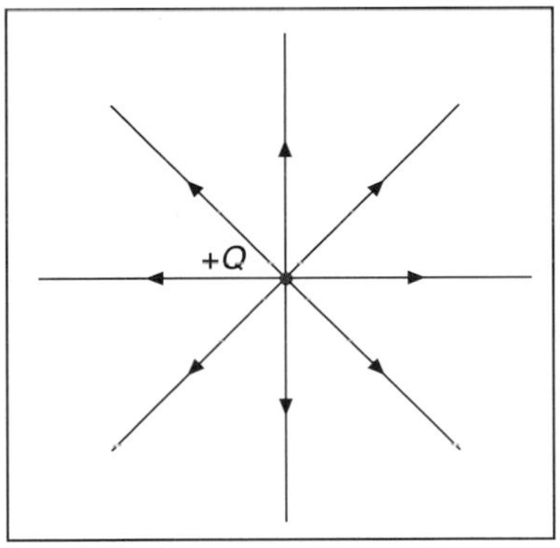

Figure 6.4 The field of a point charge is radial

Circular motion under electrostatic forces

A body can perform circular motion in a radial electrostatic field. Chapter 21 of *Waves and Our Universe* discussed one model of the hydrogen atom, which considers it to be an electron orbiting a proton (Figure 6.5).

The electrostatic attraction of the proton on the electron provides the centripetal force for the electron's motion, so

centripetal force = electrostatic force:

$$\frac{mv^2}{r} = \frac{kqQ}{r^2} \qquad \text{so} \qquad mv^2 = \frac{kqQ}{r}$$

$$\text{and} \quad v^2 = \frac{kqQ}{mr}$$

For an electron orbiting a proton, the mass m of the electron is 9.1×10^{-31} kg, and the charges q and Q are $\pm 1.6 \times 10^{-19}$ C. If the distance between their centres is approximately 0.11 nm, then

$$v = \sqrt{\left(\frac{kqQ}{mr}\right)} = \sqrt{\left(\frac{9.0 \times 10^9 \text{ N m}^2 \text{ C}^{-2} \times (1.6 \times 10^{-19} \text{ C})^2}{9.1 \times 10^{-31} \text{ kg} \times 0.11 \times 10^{-9} \text{ m}}\right)}$$

$$= 1.6 \times 10^6 \text{ m s}^{-1}$$

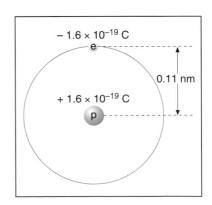

Figure 6.5 A model of a hydrogen atom

7 Uniform electric fields

Finding the shape of electric fields

Connect a 5000 V power supply to two straight electrodes immersed in castor oil, as shown in Figure 7.1. Then sprinkle semolina grains on to the castor oil. Draw the pattern that results.

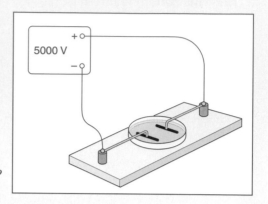

Figure 7.1 The power supply sets up an electric field between the plates

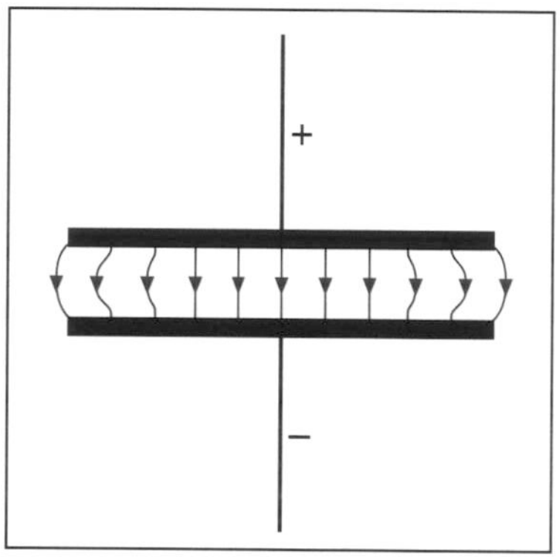

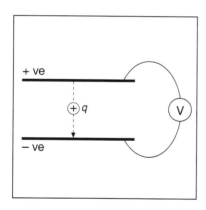

Figure 7.2 The field is uniform where the field lines are parallel

A uniform electric field

In the middle of the region between two parallel plates, the electric field lines are parallel (Figure 7.2). The lines are a constant distance apart; the electric field strength varies little in this region. This is a uniform field.

Work done in an electric field

Figure 7.3 shows a charge q between two parallel plates. When the charge moves from one plate to the other, the work done = force × distance = Fx.

The energy of the system changes as the charge moves from one plate to the other. You know from Chapter 7 of *Electricity and Thermal Physics* that change in energy = charge × potential difference. When the charge q in Figure 7.3 moves from one plate to the other, the change in energy is qV.

Figure 7.3 When charge moves between the plates, work done = qV

Equipotentials in a uniform field

- Connect a 2 V a.c. power supply to a pair of straight copper electrodes in copper sulphate solution. Use a digital voltmeter to check the voltage across the electrodes. Then put the probe in the copper sulphate solution and find places where the potential is 1 V (Figure 7.4).
- Sketch a diagram to mark the arrangement of these places. Repeat for voltages of 0.5 V and 1.5 V.

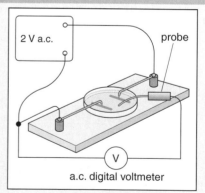

Figure 7.4 Find the places where the voltage is 1 V

Equipotential surfaces

Figure 7.5 shows two parallel plates at potentials (voltages) 0 V and 2 V. The lines between them show the potentials at places between these plates. As you would expect, the potential drops steadily from 2 V to 0 V across the gap between the plates. The lines of equal voltage on the diagram are **equipotentials**. In the uniform field between the two plates, the equipotentials are evenly spaced.

'Equipotential' means equal energy. (Strictly it means equal energy per unit charge, since potential is energy per unit charge.) If a charge moves along an equipotential, its energy does not change.

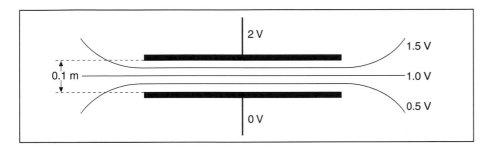

Figure 7.5 Equipotentials between two parallel plates

Compare Figure 7.5 with Figure 7.2, which shows the electric field lines between two parallel plates. You can see that the field lines are perpendicular to the equipotentials.

Potential gradient

Figure 7.6 shows a graph of potential against distance across the centre of the plates of Figure 7.5. The graph is a straight line: the potential drops uniformly with distance.

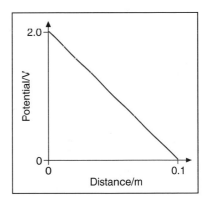

Figure 7.6 Potential–distance graph for Figure 7.5

Figure 7.7 shows a potential–distance graph for two plates a distance x apart with a potential difference V between the plates. A positive charge between the two plates would be repelled from the positive plate and attracted to the negative plate. It moves down the potential gradient.

You know that, when a charge moves from one plate to another, the work done $= qV$. This is equal to force × distance $= Fx$. So

$$Fx = qV \qquad \text{or} \qquad F/q = V/x$$

F/q is the electric field strength, E. So $E = V/x$. V/x is the **potential gradient**. It is the electrical 'slope' of a field. In uniform fields, potential–distance graphs are straight lines. The potential gradient is constant and the electric field strength is constant. When the potential gradient is large, the electrical slope is large and the electric field strength is large, so there is a large force on charged particles.

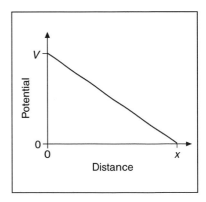

Figure 7.7 The electric field strength equals the potential gradient: $E = V/x$

8 Comparing gravitational and electric fields

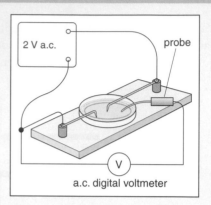

Equipotentials in a radial field

- Connect a 2 V a.c. power supply to a circular and a point electrode in copper sulphate solution. Use a digital voltmeter to check the voltage across the electrodes. Then put the probe in the copper sulphate solution and find places where the potential is 1 V (Figure 8.1).
- Sketch a diagram to mark the arrangement of these places. Repeat for voltages of 0.5 V and 1.5 V.

Figure 8.1 Investigate the voltages between the centre and the outside circle

Equipotentials for a radial field

Figure 8.2 shows the equipotentials for a radial field. The potential gets less the further you are from the charge, and the equipotentials get further apart. The electric field strength, the potential gradient, is weaker at a greater distance from the charge.

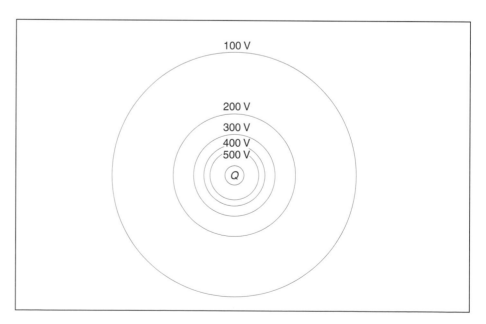

Figure 8.2 The equipotentials around a point charge Q are further apart where the field is weaker

Gravitational potential

Electrical potential is measured in volts. It is the energy per unit charge. (Remember the volt is a joule per coloumb.) Electric equipotential lines are lines of equal energy for a charge. In the same way you can draw diagrams to show gravitational equipotentials – lines of equal gravitational potential energy.

Figure 8.3 shows gravitational equipotential lines around the Earth. As you would expect, these lines are circles, equal distances from the centre of the Earth. Anywhere along the lines, a mass has the same energy.

The lines in Figure 8.3 are labelled with zero energy at the surface of the Earth. You can use them to find how much energy it needs to move 1 kg from the surface of the Earth to that line. As with electrical potential, the equipotential lines get further apart the further you are from the Earth.

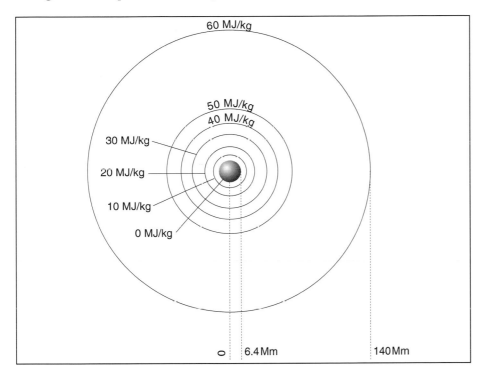

Figure 8.3   About 60 MJ is needed to take each kilogram a third of the way to the Moon

Escape speed

To get 1 kg completely free from the Earth needs 63 MJ, only 3 MJ more than the energy needed to get 1 kg to the limits of Figure 8.3 – a third of the way to the Moon. If you give 63 MJ of kinetic energy to a mass of 1 kg, ignoring air resistance, it has enough energy to become completely free of the Earth's surface. Such an object is travelling at its **escape speed**.

So $\frac{1}{2}mv^2 = \frac{1}{2} \times 1\,\text{kg} \times v^2 = 63\,\text{MJ}$

$\therefore v^2 = 2 \times 63\,\text{MJ}/1\text{kg} = 126 \times 10^6\,\text{m}^2\,\text{s}^{-2}$

$\therefore v = 11.2\,\text{km s}^{-1}$

This is about 33 times the speed of sound in air.

COMPARING GRAVITATIONAL AND ELECTRIC FIELDS

There are many similarities between gravitational fields, and a few differences. The table below makes some comparisons between the two types of field.

Gravitational fields	Electric fields
Gravitational fields affect all masses	Electric fields affect all charges
Gravitational forces are proportional to mass: $$F=mg$$	Electric forces are proportional to charge $$F=QE$$
The gravitational field strength is the force per unit mass. Its unit is the newton per kilogram.	The electric field strength is the force per unit charge. Its unit is the newton per coulomb.
Gravitational equipotentials are lines of equal gravitational energy	Electrical equipotentials are lines of equal electrical energy.
The forces between point masses obeys an inverse square law: $$F=\frac{Gm_1m_2}{r^2}$$ This is Newton's law of gravitation. Notice how similar it is to Coulomb's law.	The forces between point charges obeys an inverse square law: $$F=\frac{kQ_1Q_2}{r^2}$$ This is Coulomb's law. Notice how similar it is to Newton's law of gravitation.
All masses attract each other. There are no repulsive forces with gravity.	All like charges repel. Unlike charges attract. Electric fields have both attractive and repulsive forces.
Point masses, and spherical masses, produce a radial gravitational field.	Point charges, and spherical charges, produce a radial electric field.
Near to a spherical body, there is a uniform gravitational field.	Near to a spherical charge, or between two parallel plates, there is a uniform electric field.
There is nothing you can put round a mass to shield other masses from its gravitatiional effect.	You can shield a charge so that it has no effect on another charge. If you put a metal container connected to Earth around the charge, you shield other charges from its effect.

Capacitors

Capacitors store energy. In a power supply, capacitors charge up and discharge many times per second to help provide a steady voltage from an alternating source.

The two large smoothing capacitors (blue cases) help to produce a steady voltage from the alternating voltage provided by the power supply's transformer.

Charging a capacitor

Charging a reservoir

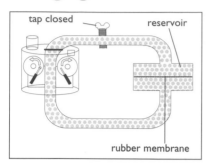

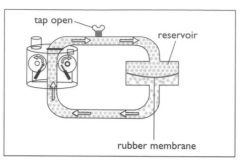

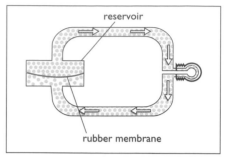

Figure 9.1a shows a reservoir of balls that can store energy. With the tap shut, the balls cannot move and the ball reservoir is uncharged.

Figure 9.1b When the tap is open the engine pushes balls into the top half of the reservoir, the rubber membrane stretches, and this pushes balls out of the bottom half. The charged reservoir doesn't really store balls; it just has more balls in the top and fewer balls in the bottom.

Figure 9.1c If you connect the charged reservoir to a load, the membrane will push balls from the top part of the reservoir to the bottom part, powering the load for a short time.

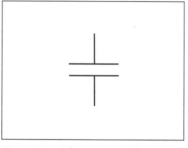

Figure 9.2 *The circuit symbol for a capacitor*

An electrical reservoir

A **capacitor** is an electrical reservoir. It consists of two metal plates separated from each other by an insulating material. A wire from each plate connects the capacitor to the circuit. The circuit symbol shown in Figure 9.2 represents this structure. In practice, the metal plates are usually made of thin foil. They are often made in long strips with a thin insulator in between and rolled up into a cylinder resembling something like a Swiss-roll.

Monitoring the current in a capacitor circuit

- Connect a 1000 μF capacitor in series with a resistor and a battery pack, and a microammeter to record the current flow (Figure 9.3).
- Use a wire to short out the capacitor as shown by the dotted line. This effectively removes the capacitor from the circuit. Record the values of current, voltage and resistance and check that they agree with each other.
- Now remove one end of the shorting lead and observe the reading on the ammeter. Record the current every 5 s, starting from when the shorting lead was removed until the current has fallen to 5 per cent of its starting value. Plot a graph of current against time.
- Repeat the experiment with the microammeter connected at point B in the circuit. Plot another graph on the same axes.
- Now change the charging resistance, first to 220 kΩ and then to 470 kΩ. For each resistor, repeat the experiment.

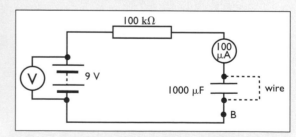

Figure 9.3 *Measuring the current in a capacitor circuit*

Current in a capacitor circuit

You might be surprised that any current can flow in a series circuit containing a capacitor. After all, there is an insulating gap between the two plates and certainly no charge can flow through it. But the capacitor works like the ball reservoir. The battery draws charge from one plate and transfers an equal charge to the other plate without any charge going directly from one plate to the other (Figure 9.4).

At any instant the current through each lead of the capacitor is the same. This shows that an equal quantity of charge is going on to one of the capacitor's plates as is leaving the other.

A charged capacitor has extra charge on one plate and an equal shortage on the other. Just like the number of balls in the reservoirs in Figure 9.1, the total charge within the capacitor remains the same. Charge is not actually stored, but moved from one place to another.

Why does the current change with time?

Kirchhoff's second law tells you that the voltage across the battery is equal to the voltage across the resistor plus the voltage across the capacitor:

$$V_{battery} = V_{resistor} + V_{capacitor}$$

At the beginning, when the capacitor is uncharged, the voltage across its plates is zero, so the voltage across the resistor is equal to the voltage across the battery. The battery voltage is constant. As the capacitor charges, the voltage across it increases, and the voltage across the resistor decreases. This means that the current through the resistor, and therefore the whole circuit, decreases (Figure 9.5). Eventually, when the capacitor is fully charged, the voltage across the capacitor is equal to the battery voltage. There is no voltage across the resistor and so no current in the circuit.

Charging quickly or slowly

At the beginning, the capacitor is uncharged and has no voltage across it. So all the battery voltage is across the resistor. Therefore the initial current is:

$$I_{max} = \frac{V_{resistor}}{R} = \frac{V_{battery}}{R}$$

As the capacitor charges up, the voltage across the resistor decreases.

Figure 9.6 shows three different graphs using the same battery and capacitor with three different charging resistances. Notice how the charging current is greater and the charging time is less for a smaller resistance.

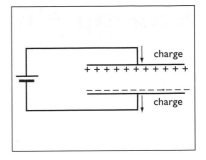

Figure 9.4   You can describe charging the capacitor in terms of the flow of positive charge

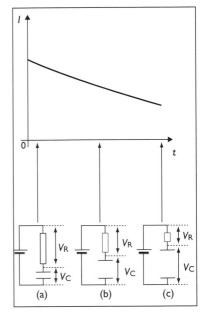

Figure 9.5   Change of current with time when charging a capacitor

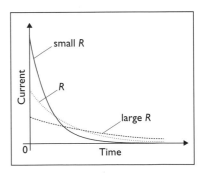

Figure 9.6   If you decrease the resistance in the charging circuit, you will increase the charging current. It then takes less time for the same charge to flow

10 Storing charge

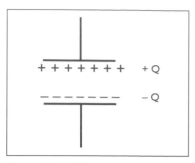

Figure 10.1   When a capacitor is charged, charge is displaced from one plate to another

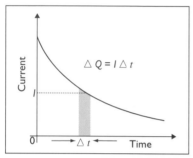

Figure 10.2   The shaded area is the charge flowing in time Δt

The meaning of capacitance

Capacitance is the ability of a component to store charge. But this can be misleading. The total charge at all times on a capacitor is zero; when it is charged, it holds equal amounts of positive and negative charges. Figure 10.1 shows a capacitor in which a charge Q is displaced from one plate to the other. The charge stored is said to be Q. As you can see, this means that the charge on the positive plate is $+Q$ and the charge on the negative plate is $-Q$. The quantity of charge Q that is displaced determines the value of its capacitance.

Finding the quantity of charge on each plate

Figure 10.2 shows a current–time graph for a capacitor charging. When a current I flows onto a capacitor for a short time Δt, the charge $\Delta Q = I\Delta t$. This is the shaded area of the graph.

The total charge that flows is the total area under the curve. This is true for any current–time graph. You can find the charge stored by a capacitor by plotting a current–time graph and determining the area under the curve.

Measuring the area under a current–time graph

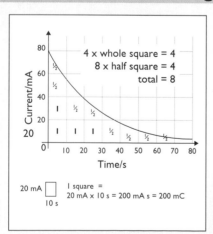

- Use the experiment 'Monitoring the current in a capacitor circuit' in Chapter 9 to plot a current–time graph for a capacitor of 1000 μF charged to 3 V through a 100 kΩ resistor.
- Determine the charge that flows by finding the area under the curve as is done in Figure 10.3.
- Repeat the charging experiment for 4.5 V and 6 V.
- For each experiment, divide the charge by the voltage and compare your figures.

Figure 10.3   Calculating the area under a current–time graph: first choose a convenient size of square to count up the area; count up the number of squares, calling each significant part-square a half; calculate the charge represented by one square, and multiply this by the number of squares to find the total charge

The definition of capacitance

The charge displaced Q is directly proportional to the charging potential difference V. If you plot a graph of charge against voltage, you get a straight line through the origin (Figure 10.4). The constant of proportionality between

Q and V (and the gradient of the Q–V graph) is the capacitance, which is defined by the following equation:

$$\text{capacitance} = \frac{\text{charge displaced}}{\text{voltage across plates}}$$

$$C = \frac{Q}{V}$$

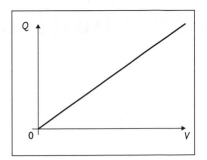

Figure 10.4   Graph of Q against V for a capacitor

From this equation, you will see that the unit of capacitance is the coulomb per volt ($C\ V^{-1}$), called the farad (F). A farad is an extremely large capacitance and most values that you meet will be given in microfarads (μF), nanofarads (nF) or picofarads (pF).

Suppose the capacitor of Figure 10.4 has a charge of 1600 μC, for a charging voltage of 6 V. Then

$$\text{capacitance} = Q/V = 1600 \ \text{μC}/6 \ \text{V} = 270 \ \text{μF, approximately}$$

Another way to find the charge on each plate

If you change the series resistance as a capacitor is being charged, you can control the charging current. A useful trick is to decrease the resistance as the capacitor charges to keep the charging current constant until the capacitor is fully charged. This produces the current–time graph shown in Figure 10.5. It is simple to calculate the charge that flows if the current is constant throughout the charging process: just multiply the current by the time. This is the same as finding the area under the rectangular current–time graph.

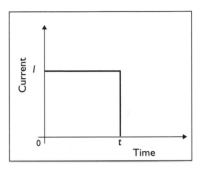

Figure 10.5   If you reduce the resistance while charging, you can keep the charging current constant

Charging a capacitor at a constant rate

- Set up the circuit in Figure 10.6. The 100 kΩ variable resistor allows you to control the charging current. Before completing the circuit, set the variable resistor to its maximum value, so that you don't pass a damagingly large current through the sensitive ammeter.
- Short the capacitor and adjust the variable resistor so that the maximum current is some convenient value on the meter.
- Remove the shorting lead and start timing; while you do so, use the variable resistor to maintain the current as near as possible to its starting value. Stop the clock when the resistor has reached its minimum value and the current drops rapidly.
- Repeat the experiment. Calculate the average time from several attempts and use this figure, together with the current and the charging voltage, to calculate the capacitance.

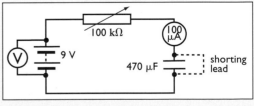

Figure 10.6   Charging a capacitor at a constant rate

ⅠⅠ Exponential decay – capacitors

Charging and discharging a capacitor

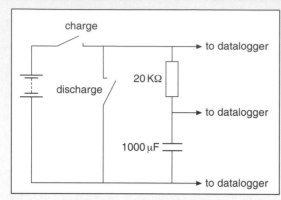

Figure 11.1 *Charge the capacitor and then discharge it*

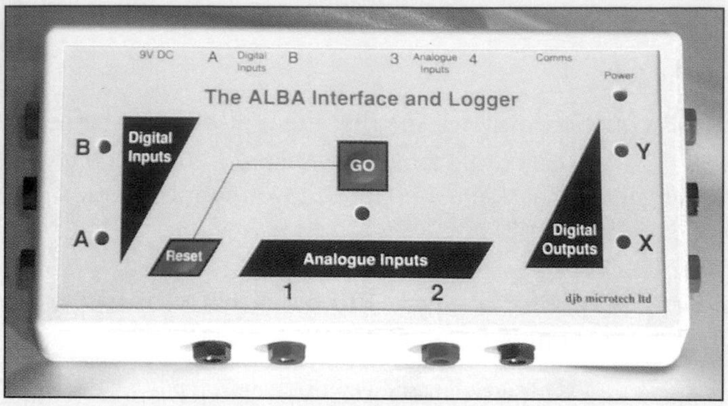

Figure 11.2 *The datalogger takes readings every second*

- Connect the circuit of Figure 11.1 to a datalogger (Figure 11.2). Record the voltages across the capacitor and across the resistor as you charge and discharge the capacitor.
- Use your results to display a current–time graph for the charging and discharging process.
- Repeat the experiment but this time use a resistance of 40 kΩ in the circuit.
- Repeat the experiment again; this time use a resistance of 20 kΩ and a capacitance of 2000 µF.

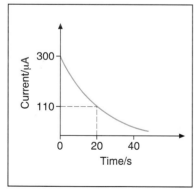

Figure 11.3 *The time constant of this discharge is 20 s*

Time constant

Both charging and discharging are like radioactive decay. It does not matter how long you observe, the process goes on indefinitely. The current never drops completely to zero. This form of decrease is called an exponential decay.

But it is still possible to define a time related to the charging or discharging graph. This is called the **time constant**. The time constant is the time for the current to decrease to 1/e of its original value. The number e is approximately 2.718. So the time constant is the time for the current to decrease to 1/2.718 of its original value. Figure 11.3 shows the charging graph for the circuit of Figure 11.1. The time constant is 20 s. During this time the current drops from 300 µA to (300 µA)/e = 110 µA.

The time constant for charging or discharging depends on the resistance and the capacitance. If you increase the resistance, the initial charging current is reduced, and the charging process takes longer. If you increase the capacitance, the amount of charge needed to charge it fully increases and, other things being equal, charging therefore takes longer. So

$$\text{time constant} = \text{resistance} \times \text{capacitance}$$
$$= RC$$

Figure 11.4 shows how the charging current varies for a circuit of time constant RC.

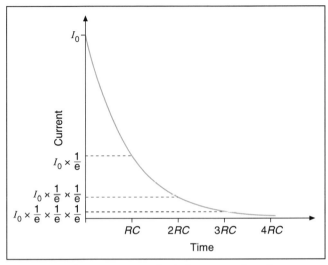

Figure 11.4 After every interval RC, the current drops by a factor 1/e

The unit of resistance is the ohm, Ω (volt per amp; see Chapter 10 of *Electricity and Thermal Physics*), and the unit of capacitance is the farad, F (coulomb per volt; see Chapter 10).

So the unit of the time constant is:

$$\Omega \times F = \frac{V}{A} \times \frac{C}{V}$$

$$= \frac{V}{A} \times \frac{As}{V}$$

$$= s$$

So, as you might expect the unit of the time constant is the second.

12 Exponential decay – radioactivity

Decay curves

Look at the discharge graphs of Chapter 11. These graphs are exponential decays. You can describe them with an exponential decay equation:

$$I = I_0\, e^{-t/RC}$$

I_0 is the initial current; I is the current at time t, and RC is the time constant. During the discharge, the voltage across the capacitor

$$V = \frac{Q}{C}$$

The current through the resistor

$$I = \frac{V}{R} = \frac{Q}{RC}$$

This shows that the discharge current I is proportional to Q the charge remaining. So you can write:

$$Q = Q_0\, e^{-t/RC}$$

where Q_0 is the initial charge on the capacitor.

Rate of decrease proportional to amount remaining

The current is the rate of flow of charge

$$I = \frac{dQ}{dt}$$

$$I = \frac{Q}{t} \quad \text{so} \quad \frac{dQ}{dt} = \frac{Q}{RC}$$

The rate of flow of charge is proportional to the charge Q remaining on the capacitor. This equation is of the same form as the radioactive decay equation

$$\frac{dN}{dt} = \lambda N$$

You can guess now that there is a similar **exponential decay** equation to describe radioactive decay.

$$N = N_0\, e^{-\lambda t}$$

The number N of undecayed atoms left after time t is proportional to the number N_0 you started with.
You can rewrite this equation as

$$N = \frac{N_0}{e^{\lambda t}}$$

26

The term $e^{\lambda t}$ increases with time because t is increasing. So the number left decreases with time. This means that dN/dt strictly is negative. If you understand differentiation, follow through the mathematics in the box on the left, which shows that

$N = N_0\, e^{-\lambda t}$ is compatible with the equation $\dfrac{dN}{dt} = -\lambda N$

> Differentiating
> $$N = N_0\, e^{-\lambda t}$$
> we get
> $$dN/dt = (N_0)(-\lambda)\, e^{-\lambda t}$$
> $$= -\lambda N_0\, e^{-\lambda t}$$
> $$dN/dt = -\lambda N$$

Straight line graph from the decay equation

If you take natural logarithms of both sides of the equation $N = N_0\, e^{-\lambda t}$, you get

$$\ln N = \ln N_0 - \lambda t \qquad \text{or} \qquad \ln N = -\lambda t + \ln N_0$$

This is of the form $y = mx + c$. So a graph of $\ln N$ against t will have slope $-\lambda$.

The connection between the half-life and the decay constant is:

$$t_{1/2} = (\ln 2)/\lambda$$

Figure 12.1 is a graph of $\ln N$ against t for a decay experiment with protactinium. We can find the half-life of protactinium as follows:

slope $= -\lambda = -6.0/(625\text{ s})$

and therefore

$$\lambda = 9.68 \times 10^{-3}\ \text{s}^{-1}$$

which gives a half-life of

$$t_{1/2} = (\ln 2)/\lambda = 0.693/(9.68 \times 10^{-3}\ \text{s}^{-1}) = 72\text{ s}$$

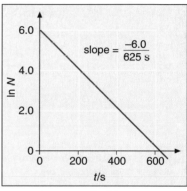

Figure 12.1   Graph of ln N against t

Half-life for capacitor discharge

The normal way of measuring the time of capacitor discharge is by the time constant RC. This is the time for the charge (or current) to decrease to $1/e$ of its original value. You can think of it as the e'th life. But you can calculate the half-life of a capacitor discharge by analogy with the radioactive decay equation.

For radioactive decay, $dN/dt = \lambda N$.
The half-life $t_{1/2} = \ln 2/\lambda$

For capacitor discharge, $dQ/dt = Q/RC$
i.e. $RC \equiv 1/\lambda$

So $t_{1/2} = \ln 2 \times RC$

Table 12.1 *Comparison between radioactive decay and capacitor discharge*

Capacitor discharge	Radioactive decay
Charge is decreasing	Number of atoms is decreasing
Rate of decrease of charge is current	Rate of decrease of number of atoms is the activity
Current $I = dQ/dt$	Activity $= dN/dt$
$dQ/dt = Q/RC$	$dN/dt = \lambda N$
$Q = Q_0\, e^{-t/RC}$	$N = N_0\, e^{-\lambda t}$
Time constant: RC Half-life: $\ln 2 \times RC$	Time constant: $1/\lambda$ Half-life: $\ln 2/\lambda$

Storing energy – capacitors and springs

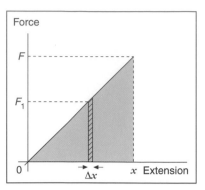

Figure 13.1 *The energy stored in the spring is the area under the force–extension line.*

Energy stored in a spring

In *Mechanics and Radioactivity* you learned about springs. The force F needed to stretch a spring an extension x is given by

$$F = k\,x$$

where k is the spring constant. Figure 13.1 shows the graph of this equation.

The work done stretching a spring the small distance Δx is the force F_1 multiplied by the distance moved in the direction of the force Δx. This is equal to $F\Delta x$, the area of the small strip under the line in Figure 13.1. The total work done stretching the spring is the total area under the line. This is $\frac{1}{2}Fx$. This is the energy stored in the spring

How much energy can a capacitor store?

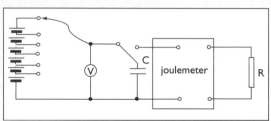

Figure 13.2 *Use a joulemeter to measure how much energy a capacitor can store*

- Set up the circuit in Figure 13.2.
- Use six cells to charge your capacitor. Note the charging voltage and then discharge the capacitor through the resistor, using the joulemeter to measure the energy transferred.
- Repeat with a range of charging voltages.
- Plot a graph of energy against voltage squared and comment on the graph.
- Repeat the experiment with twice the capacitance.

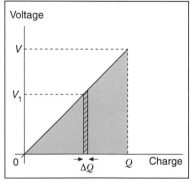

Figure 13.3 *The energy stored in the capacitor is the area under the voltage–charge line.*

Energy stored in a capacitor

For a spring, force is the quantity that causes the spring to extend. For a capacitor, voltage is the quantity that causes charge to be displaced.

For a capacitor, $Q = CV$, so $V = \frac{1}{C}Q$

You can compare this with the spring formula $F = kx$.

Figure 13.3 shows a graph of V against Q. The graph is a straight line. Its slope is $\frac{1}{C}$.

Remember that voltage is the work done per unit charge, measured in joules per coulomb. So when a power supply adds a small amount of charge ΔQ to a capacitor at a voltage V, then the work done is $V\Delta Q$. This is equal to the area of the small strip on the graph.

The total work done charging the capacitor is equal to the areas of all the vertical strips, the total area under the graph. Since the graph is a straight line, the total work done W is simply the area of a triangle and is given by the equation

$$W = \frac{1}{2}QV$$

You can combine this equation with that which defines capacitance ($C = Q/V$) to produce two alternative expressions:

$$W = \tfrac{1}{2}CV^2 \quad \text{and} \quad W = \tfrac{1}{2}Q^2/C$$

The work done by a supply as it puts a charge Q onto a capacitor is QV. The capacitor stores $\tfrac{1}{2}QV$; the remainder is dissipated in the circuit resistances as the capacitor charges. When the capacitor discharges, the $\tfrac{1}{2}QV$ that is stored is dissipated in the circuit through which the capacitor discharges. You should not be surprised that the same amount of energy is dissipated on charging and discharging. If both take place through the same resistance, both result from the charging current being the same as the discharging current and flowing for the same time, but in the opposite direction.

Table 13.1 *Comparing capacitors and springs*

Capacitors	Springs
Voltage displaces the charge	Force causes the extension
The amount of charge displaced is proportional to the voltage	The amount of displacement is proportional to the force
$V = \tfrac{1}{C}Q$	$F = kx$
Work done = average voltage × charge $= \tfrac{1}{2}VQ$	Work done = average force × distance $= \tfrac{1}{2}Fx$

Making small capacitors

Miniature electronic devices need large values of capacitance in as small a space as possible. So capacitors are built with very small plate spacing, very thin plates and with insulators of high permittivity. They are often rolled up to save space. Figure 13.4 shows a polystyrene capacitor and a diagram of its construction.

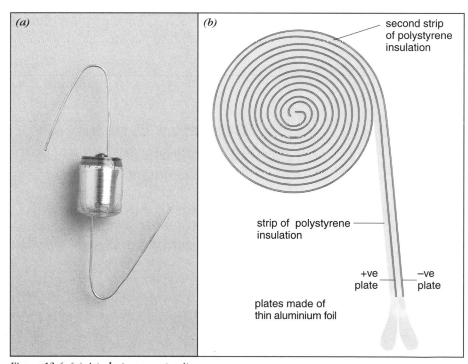

Figure 13.4 (a) A polystyrene capacitor.
(b) The plates of this small capacitor are made of thin foil which is then rolled up

14 Capacitor combinations

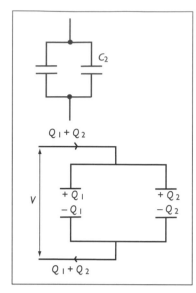

Figure 14.1   *When capacitors are connected in parallel, the total charge that flows is equal to the sum of the charges for each capacitor*

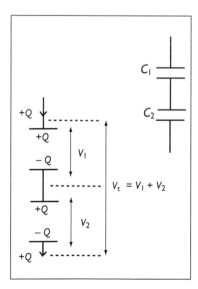

Figure 14.2   *When capacitors are connected in series, the voltage across the whole is equal to the sum of the voltages across each capacitor*

Capacitors in parallel

When you connect capacitors in parallel (Figure 14.1), the capacitances add up. The voltage across each capacitor is the same and the amount of charge on the plates of each capacitor is determined by its capacitance.

The single capacitor that could replace this arrangement would have to hold an amount of charge on its plates equal to that held by both capacitors when it has the same potential difference across it. So

$$Q_t = Q_1 + Q_2$$

But since $Q_1 = C_1 V$ and $Q_2 = C_2 V$ we can write

$$\text{total charge} = C_1 V + C_2 V$$
$$= V(C_1 + C_2)$$

and

$$\frac{\text{total charge}}{V} = C_1 + C_2$$

This represents the capacitance that is equivalent to the parallel arrangement shown:

$$C_t = C_1 + C_2$$

So if a 100 μF is connected in parallel with a 200 μF capacitor,

$$C_t = C_1 + C_2 = 100 \ \mu F + 200 \ \mu F = 300 \ \mu F$$

Capacitors in series

When you connect capacitors in series, the total capacitance is *less* than each individual capacitance.

The capacitors in Figure 14.2 form part of a series circuit. At any instant, the current in all the connecting leads must be the same. You would get the same current–time graph wherever you connected an ammeter into the circuit. Since $\Delta Q = I\Delta t$, each capacitor must have the same displaced charge. The voltage across each capacitor is determined by its capacitance.

The single capacitor that could replace this arrangement would have to hold this amount of charge on its plates when it has the total potential difference of the individual capacitors across it. So

$$V_t = V_1 + V_2$$

But since $V_1 = Q/C_1$ and $V_2 = Q/C_2$ we can write

$$\frac{Q}{C_t} = \frac{Q}{C_1} + \frac{Q}{C_2}$$

and

$$\frac{1}{C_t} = \frac{1}{C_1} + \frac{1}{C_2}$$

If you connect a capacitance of 100 μF in parallel with a capacitance of 200 μF,

$$\frac{1}{C_t} = \frac{1}{C_1} + \frac{1}{C_2}$$

$$\frac{1}{C_t} = \frac{1}{100 \ \mu F} + \frac{1}{200 \ \mu F}$$

$$\frac{1}{C_t} = \frac{200 + 100}{100 \times 200 \ \mu F}$$

$$\frac{1}{C_t} = \frac{300}{100 \times 200 \ \mu F}$$

$$C_t = \frac{100 \times 200 \ \mu F}{300} = 67 \mu F$$

WORKED EXAMPLE

Complete the circuit below to show:
the capacitors connected in parallel.

the capacitors connected in series.

Use the information in the diagrams to complete the following table.

Capacitors in parallel	Charge on C_1	$Q_1 = C_1 V = 3 \ \mu F \times 6V = 18 \ \mu C$
	Energy stored in C_1 when fully charged	$W_1 = \frac{1}{2} C_1 V^2 = \frac{1}{2} \times 3 \ \mu F \times 36 \ V^2$ $= 54 \ \mu J$
Capacitors in series	Charge on C_2	Voltage across $C_2 = \frac{1}{2} \times 6 \ V = 3 \ V$ $Q_2 = C_2 V_2 = 3 \ \mu F \times 3 \ V = 9 \ \mu C$
	Work done by power supply as it charges both capacitors	9 μC flows through 6 V supply $W = Q V = 9 \ \mu C \times 6 \ V = 54 \ \mu J$

CAPACITOR COMBINATIONS

Table 14.1 *Comparing resistors and capacitors*

Resistors	Capacitors
In series $$R_\mathrm{t} = R_1 + R_2$$	In series $$\frac{1}{C\mathrm{t}} = \frac{1}{C_1} + \frac{1}{C_2}$$
In parallel $$\frac{1}{R_\mathrm{t}} = \frac{1}{R_1} + \frac{1}{R_2}$$	In parallel $$C_\mathrm{t} = C_1 + C_2$$

Magnetism

Albert Einstein showed that magnetic forces are *electrostatic* forces viewed through the eyes of relativity. Magnetism is caused by the *movement of charge* – whether it is the rotation of electrons around the nucleus of an atom, or the flow of electrons in the wires of an electromagnet.

Magnetism levitation – a cylindrical magnet 'floats' above a ceramic superconductor cooled by liquid nitrogen.

Magnets

Bar magnets

- Place a bar magnet on a cork floating in water. Compare its final orientation with that of a compass needle placed 1 m away.
- What happens as you bring the compass needle nearer and nearer to the floating magnet?
- Hold a bar magnet in each hand. Feel the force exerted as you slowly bring the ends of the magnets together. Reverse one magnet and repeat. What happens if you now reverse the other magnet?

Figure 15.1 Suspended magnets align north–south

Magnetic fields

Any freely suspended bar magnet will rotate until it points in a north–south direction, as shown in Figure 15.1.

The end of the bar magnet pointing towards the Earth's magnetic North Pole is referred to as its north-seeking pole or, more simply, its north pole. The other end is its south-seeking, or south, pole.

As the north poles of two bar magnets are brought together, they repel. The same thing happens with two south poles. A north and a south pole attract.

A force acts between the poles before they are touching. This force increases as the poles get closer together. Around a bar magnet there is a **magnetic field**, a region within which there are magnetic forces.

A suspended magnet aligns with the Earth's magnetic field unless there is a stronger magnetic field present. Two compass needles held close together affect each other, so that neither points towards magnetic north.

Magnetic field patterns

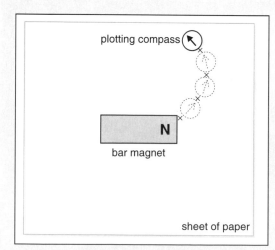

Figure 15.2 Plotting a magnetic field pattern

- Support a piece of stiff card above a bar magnet. Sprinkle some iron filings onto the card and tap it gently. Sketch the observed pattern.
- Investigate the field between two magnets, first with opposite poles facing and then with like poles facing.
- Place a bar magnet in the centre of a sheet of white paper, trace its outline and put a cross close to its north pole. Place a small plotting compass so that the south end of its needle is above the cross. Mark the position of the north end of the needle. Continue to move the plotting compass, marking the position of its north end (Figure 15.2).
- Repeat this procedure for various starting points around the magnet.

Magnetic field lines

Iron filings show the shape of a magnetic field by aligning themselves with the magnetic field lines. The north end of the needle of a plotting compass indicates the direction of a magnetic field. Magnetic field lines leave the north pole of a magnet and enter at its south pole, as shown in Figure 15.3. The field lines are closest together at the poles where the magnetic field is strongest.

The magnetic field lines between two attracting poles go from one pole to the other, seeming to pull the magnets together.

The magnetic field lines between two like poles do not combine, as the poles push each other apart (Figure 15.4).

Neutral points

A **neutral point** is a position within overlapping magnetic fields where the fields cancel so that the resultant field is zero. A neutral point exists between the two repelling magnets in Figure 15.4 where their magnetic fields cancel.

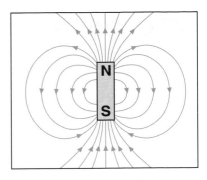

Figure 15.3 Magnetic field of a bar magnet

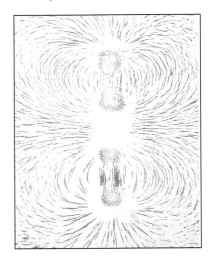

Figure 15.4 The two bar magnets beneath this card are repelling

Investigating neutral points

- Place a magnet North–South as shown in Figure 15.5. The resultant field is as shown in the diagram.
- Use a plotting compass to find the neutral points and measure how far they are from the magnet.
- Repeat but place the magnet with its North and South poles reversed.
  Sketch the resultant field diagram.
- Repeat again with the magnet East–West.

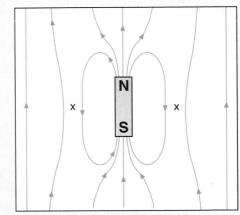

Figure 15.5 At the neutral points, X, the Earth's magnetic field and the field due to the bar magnet cancel each other

Magnetic effects of currents

The magnetic field of a current-carrying wire

- Push a length of insulated copper wire vertically through the centre of a horizontal piece of card (Figure 16.1). Pass a current of 5 A through the wire, noting the direction of the current.
- Use iron filings and a plotting compass to investigate the magnetic field around the wire.

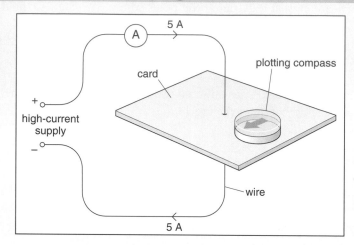

Figure 16.1 The space around a wire carrying a current contains a magnetic field

The corkscrew rule

Any moving charge produces a magnetic field. The magnetic field lines due to a current through a wire are circles centred on the wire. A plotting compass shows that the field is clockwise when viewed along the direction of current flow. The **corkscrew rule** is a useful way of remembering this: a corkscrew has to be rotated clockwise to drive it forward into the cork of a wine bottle.

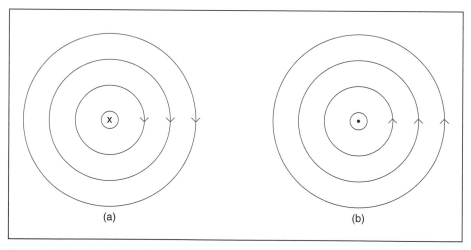

Figure 16.2 Magnetic field of a current flowing (a) into the page and (b) out of the page

Figure 16.2a shows the magnetic field looking down the wire in the direction of the current flow; the cross at the centre of the diagram is the symbol for current flowing into the diagram; you can imagine it as the back end of a dart carrying current away from you. Figure 16.2b shows the field due to a current coming out of the diagram; you can imagine the dot at the centre as the point of a dart carrying current towards you.

The magnetic field of a current-carrying cylindrical coil

- Make a solenoid (a long coil) about 10 cm long with 20 turns of wire. Pass a current of 5 A through it (Figure 16.3).
- Sketch the shape and direction of the magnetic field produced by the solenoid.
- Compare the current directions in the wire where they pass through the holes in the card. Sketch the shape and direction of the magnetic field produced by each wire.

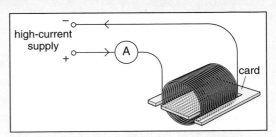

Figure 16.3 Current through coils

Magnetic fields associated with coils

The magnetic field associated with a current flowing in a coil is a resultant of all the magnetic fields produced around each part of each individual coil of wire.

A **solenoid** is a cylindrical current-carrying coil of wire with a large number of turns. A section through its magnetic field pattern is shown in Figure 16.4. The magnetic field lines enter the solenoid at one end, leave at the other end, and loop around the outside to re-enter the solenoid at the same point. The field lines are continuous.

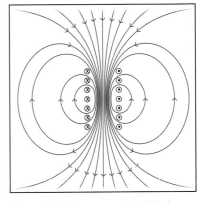

Figure 16.4 The magnetic field of a solenoid is like a bar magnet

Through the centre of the solenoid, the magnetic fields produced by each of the current rings are all in the same direction, producing a larger resultant magnetic field along its axis. When a solenoid is long and narrow, the field lines through its centre are parallel and equally spaced, indicating that the magnetic field is uniform in this region.

Compare the magnetic field pattern of a solenoid (Figure 16.4) with that of a bar magnet (Figure 15.3). The shapes of the 'outside' fields are similar. Indeed, a common method of producing a bar magnet is to place a suitable length of steel inside a solenoid carrying a direct current. The field lines through the centre of the solenoid are like the aligned domains within the bar magnet.

The ends of a solenoid behave like the poles of a bar magnet, the north pole being the end where the field lines leave. Figure 16.5 shows a simple way of telling which solenoid pole you are facing from the direction of current flow.

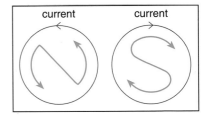

Figure 16.5 If you look at the end of a solenoid, the current direction shows which pole it is

Fleming's left-hand rule

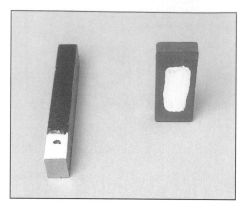

Figure 17.1 Bar and magnadur magnets

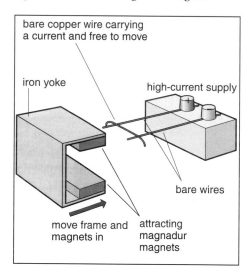

bare copper wire carrying a current and free to move

iron yoke

high-current supply

bare wires

move frame and magnets in

attracting magnadur magnets

Figure 17.2 Position the attracting magnets above and below the wire

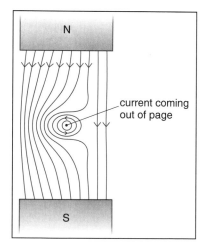

N

current coming out of page

S

Figure 17.3 Resultant of uniform and circular magnetic fields

Magnadur magnets

Figure 17.1 shows two magnets with their north poles painted white. The bar magnet is long and thin with a pole at each end. The magnadur magnet is short and fat. Its poles are the large faces. A strong and uniform magnetic field is produced in the space between two attracting magnadur magnets held close together.

Magnetic catapult

- Place two attracting magnadur magnets on opposite sides of a soft-iron, U-shaped yoke. Use a plotting compass to find the direction of the magnetic field between them.
- Connect two parallel, horizontal lengths of stiff, bare wire to a power supply. Place a third length of bare wire across the other two, completing the circuit. Hold the yoke with the magnets above and below the wire (Figure 17.2) with their magnetic field acting downwards.
- What happens to the free wire when the power supply is turned on?
- Repeat with the magnetic field of the magnets acting upwards.
- What happens in each case if the current is reversed?

Catapult field

A wire carrying a current has its own magnetic field. When such a wire is in another field, the effect of the two fields is to produce a force on the wire. Figure 17.3 shows the magnetic field that is produced when a wire carries a current perpendicular to a uniform magnetic field.

To the right of the wire, the two magnetic fields are in the opposite direction and cancel out, while to its left they are in the same direction and reinforce. The resultant magnetic field is much stronger on the left, and the field lines get distorted around to this side of the wire. Think of the field lines as lengths of elastic, each one trying to return to the shortest possible length. It looks as though they are pushing the wire towards the right, and indeed there is a force on the wire in that direction.

The field due to the magnets is downwards; the current is out of the paper; the resulting force is to the right. All these three directions are at right angles to each other.

Fleming's left-hand rule (Figure 17.4) provides a way of remembering these directions, and can be used to predict the direction of one of these when the other two are known.

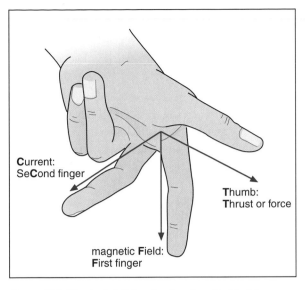

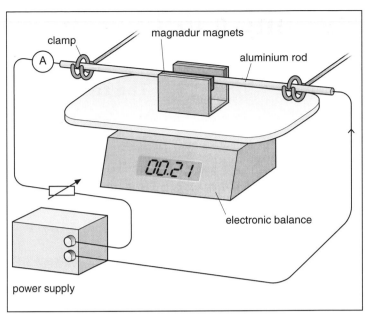

Figure 17.4 Fleming's left-hand rule: when the First finger points in the direction of the magnetic Field and the seCond finger points in the direction of the Current, the Thumb gives the direction of the Thrust (or force) on the conductor

Figure 17.5 The electronic balance measures the force produced by the interaction of the two magnetic fields

Force on a current-carrying conductor in a magnetic field

- Set up the apparatus shown in Figure 17.5 as follows. Place the yoke with two attracting magnadurs on the balance. Zero the balance. Clamp an aluminium rod horizontally between the poles of the magnets. Pass a current of 1 A through the rod.
- Record the balance reading. What does a negative balance reading indicate? How could such a reading be made positive? Record a set of readings of current and force.
- Place another yoke of magnadurs on the balance to double the length of rod in the magnetic field. Make sure both fields are in the same direction. Record another set of current and force readings. Repeat using three yokes.
- On the same axes, plot a graph of force against current for each set of readings. What does your graph show?
- Observe the effect of a stronger magnetic field on the force by holding the magnadur magnets nearer to the rod.

How the force varies

The balance in Figure 17.5 registers the force exerted on the yoke: a force that is equal and opposite to that exerted on the clamped rod. A positive reading indicates that the yoke is being pushed down and that there is an upward force on the rod.

Increasing the magnetic field increases the force. The graphs in Figure 17.6 show that the force on a current-carrying wire in a magnetic field is proportional to the current.

For a given current, the force is directly proportional to the length of the current-carrying wire in the magnetic field. The graphs in Figure 17.6 are twice as steep when using twice the length, and three times as steep with three times the length.

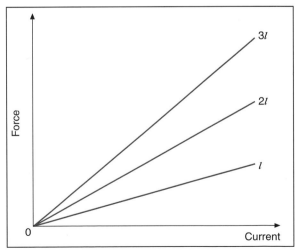

Figure 17.6 Force exerted on different currents for different lengths of conductor

Magnetic field strength

The tesla

In the last chapter, we saw that a conductor carrying a current I perpendicular to a magnetic field experiences a force F that is proportional to both the current flowing and the length l of conductor in the field. The force also depends on the strength of the magnetic field:

$$\text{force} = \text{magnetic field strength} \times \text{current} \times \text{length}$$

$$F = BIl$$

where B is the strength of the magnetic field called the **magnetic flux density**:

$$B = F/Il$$

So the SI unit of B is $\text{N A}^{-1}\,\text{m}^{-1}$. This is called the **tesla** (T). A magnetic flux density of 1 T produces a force of 1 N on each 1 m length of wire carrying a current of 1 A perpendicular to the field. Table 18.1 gives some typical values of magnetic flux density.

Table 18.1 *Typical values of magnetic flux density*

	Typical magnetic flux density/T
Earth	10^{-5}
air-cored solenoid	10^{-3}
magnadur magnet	10^{-1}
iron-cored solenoid	1

WORKED EXAMPLE

Calculate the force exerted on 8 cm of wire when it carries a current of 2 A at right angles to a field of magnetic flux density 20 mT.
We get

$$F = BIl = 20 \times 10^{-3}\,\text{T} \times 2\,\text{A} \times 8 \times 10^{-2}\,\text{m} = 0.0032\,\text{N} = 3.2\,\text{mN}$$

Finding magnetic flux densities

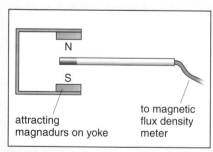

Figure 18.1 Using a Hall probe

- Repeat the experiment on the previous page. Measure the force produced by a known current flowing perpendicularly to the field produced by a pair of attracting magnadurs. Measure the length of wire in the field. Calculate the magnetic flux density, $B = F/Il$, between the attracting magnadur magnets in the yoke.
- Insert a calibrated Hall probe (Figure 18.1) between the attracting magnadur magnets and compare its reading with your value calculated from the measured force. Account for any differences.

The Hall probe

A **Hall probe** consists of a small, thin rectangle of semiconductor material. A constant current is passed through the length of the slice. When the probe is placed at right angles to a steady magnetic field, a voltage appears across the width of the slice. This Hall voltage, which is of the order of microvolts, is directly proportional to the flux density of the magnetic field. A sensitive digital voltmeter, connected across the slice, records the Hall voltage. Some Hall probes are pre-calibrated using known flux densities and give readings directly in teslas.

Measuring magnetic fields

- Insert a pre-calibrated Hall probe into the centre of a solenoid to measure the magnetic field strength there. Investigate how the field strength varies with the current flowing through the solenoid (Figure 18.2). Plot a graph of magnetic flux density against current.
- Connect three solenoids, with the same length and number of turns but different cross-sectional areas, in series. Pass a current of about 3 A through them. How does the magnetic flux density at the centre of a solenoid depend on its area?
- Two clamped half-metre rules (Figure 18.3) have 50 slinky turns between them. With the rules placed 20 cm apart, the number of turns per unit length is $50/(0.20\text{ m}) = 250\text{ m}^{-1}$. Record the flux density at the centre of this solenoid for a current of 8 A. Repeat using different numbers of turns per unit length. Plot a graph of magnetic flux density against number of turns per unit length.
- Use a Hall probe to measure the magnetic field strength 1 cm from a long straight wire carrying a current of 10 A (Figure 18.4). Repeat for a range of distances, and plot a graph of B against $1/r$.
- With r fixed at 1 cm, investigate how B depends on I.

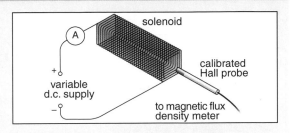

Figure 18.2 *The flux density meter will read maximum when the field lines are perpendicular to the flat end of the probe*

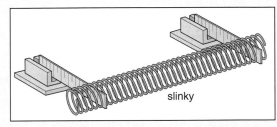

Figure 18.3 *Stretch the slinky to reduce the number of turns per unit length*

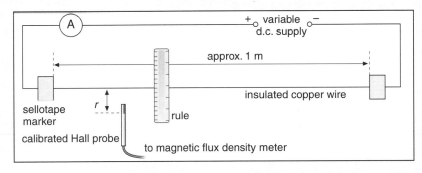

Figure 18.4 *Position the probe so that it is perpendicular to the circular magnetic field around the wire*

Permeability of free space

The magnetic field strength B within a solenoid is independent of the cross-sectional area of the solenoid, provided the length of the solenoid is large compared with its diameter. It depends on the current I flowing and on the number of turns per unit length, n (the turns density).

If the region inside a solenoid is air or a vacuum,

$$B = \mu_0 n I$$

where μ_0 is a constant called the permeability of free space and has the value $4\pi \times 10^{-7}\text{ N A}^{-2}$.

For a long straight wire, B depends on the current I and the distance r from the wire:

$$B = \frac{\mu_0 I}{2\pi r}$$

Laws of electromagnetic induction

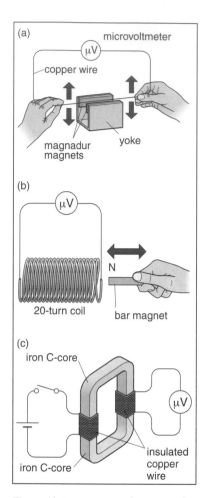

Figure 19.1 Moving conductors and magnetic fields

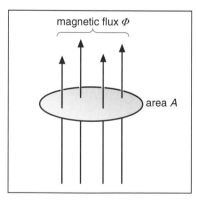

Figure 19.2 The flux density B = Φ/A

Moving conductors and magnetic fields

When there is a changing magnetic field in a circuit, or when a wire moves relative to a magnetic field, voltages are induced (generated). A voltage is induced when a conductor moves relative to a magnetic field (Figure 19.1a). Likewise, a voltage is induced when a magnetic field moves relative to a conductor (Figure 19.1b). Reversing the direction of motion of either induces a voltage in the opposite direction. A larger voltage results from faster movement, a stronger field and more turns. Figure 19.1c shows another way of moving a magnetic field relative to a conductor. What happens when the switch is closed and when it is opened?

Magnetic flux

You learnt in Chapter 18 that the strength of a magnetic field is called the magnetic flux density. This sounds as though it is to do with something flowing. You can imagine the magnetic field lines to be lines of **magnetic flux** or flow. Where the lines are closest, the flow density or flux density is greatest, and the field is strongest. With a bar magnet, the flux flows from north to south around the magnet. The flux flows in a circular path around a wire carrying a current.

The magnetic flux density B through an area is the amount of flux that passes through that area divided by the area (Figure 19.2):

$$\text{magnetic flux density} = (\text{magnetic flux})/\text{area}$$

so
$$\text{magnetic flux} = \text{magnetic flux density} \times \text{area}$$

$$\Phi = BA$$

Here Φ, the Greek capital phi ('fi', rhymes with 'pie'), is the symbol for magnetic flux.

The SI unit of magnetic flux is the tesla metre squared (T m^2), called the weber (Wb). A tesla is equal to a weber per metre squared.

When a coil having N turns surrounds a magnetic flux Φ, each turn links a flux Φ. The total **magnetic flux link** the coil is $N\Phi$ (Figure 19.3).

Explaining the induced current

As a wire moves perpendicularly through a magnetic field, conduction electrons within the wire move with the wire and can be regarded as a current flowing through the field. The electrons experience a force acting at 90° to both field and direction of motion.

You can apply Fleming's left-hand rule to these conduction electrons (Figure 19.4). Remember that electrons moving to the right form a conventional current moving to the left. The electrons are forced towards end Y of the wire, which becomes negatively charged. End X becomes positively charged. A voltage or e.m.f. is induced across the ends of the wire. If the wire forms part of a circuit, then an induced current flows.

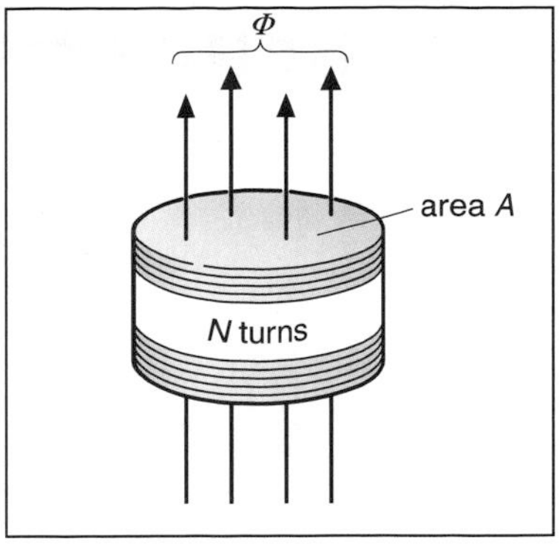

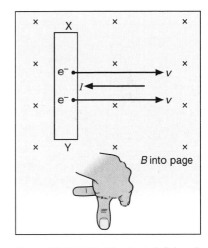

Figure 19.3 Magnetic flux linkage through coil is $N\Phi = NBA$

Faraday's law

Michael Faraday (1791–1867) discovered that the magnitude of an induced e.m.f. is directly proportional to the rate of change of magnetic flux linkage. So the induced e.m.f. $= N\dfrac{\mathrm{d}\Phi}{\mathrm{d}t}$. This is **Faraday's law**.

Direction of induced current and field

- Connect a cell, resistor and microvoltmeter to a coil of wire (Figure 19.5a). Which way does the meter deflect? Use the polarity of the cell to determine the direction of current flow around the coil. Identify the north end of the coil.
- Remove the cell and resistor from the circuit (Figure 19.5b). Thrust the north pole of a bar magnet into what was the north end of the coil. Which way does the meter now deflect? What happens when the magnet is withdrawn?

Lenz's law

If you move the north pole of a magnet towards a loop of wire, it induces a current in that loop. The direction of the current makes the end of the loop facing the magnet itself acquire a north polarity, which repels the north pole of the magnet back. If the north pole moves away from the loop, this results in the facing end of the coil acquiring a south polarity, which pulls the north pole back, opposing the removal of the magnet from the coil.

Figure 19.4 Apply Fleming's left-hand rule to the conduction electrons

This situation is an example of **Lenz's law**, which states that any current driven by an induced e.m.f. opposes the change causing it.

Faraday's law of electromagnetic induction is often written as

$$\text{induced e.m.f.} = -N\frac{\mathrm{d}\Phi}{\mathrm{d}t}.$$

The negative sign indicates that the induced e.m.f. would produce a current that opposes the change causing it.

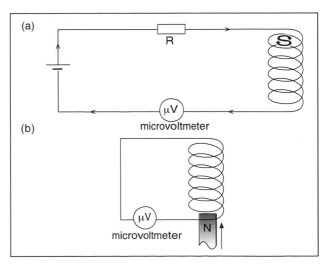

Figure 19.5 Finding the induced magnetic polarity

Applications of electromagnetic induction

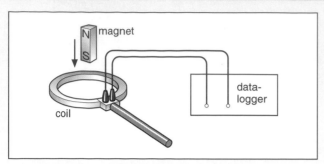

Figure 20.1 The data-logger samples the induced e.m.f. every 2 ms

- Connect a horizontal 300-turn coil to a voltage data-logger. Set the sample frequency to 500 Hz. Start the data-logger and immediately release the bar magnet (Figure 20.1).
- Use the recorded data values to plot a graph of induced e.m.f. against time. Explain the shape of the graph.
- Repeat the experiment, first reversing the magnet, and then dropping it from a greater height.

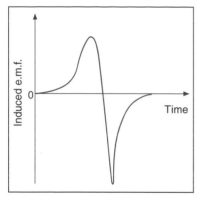

Figure 20.2 The e.m.f. induced across the coil

Changing magnetic flux linkage

If you drop a bar magnet through a coil, the flux through the coil increases as the bar magnet enters and decreases as it leaves. An e.m.f. is induced across the coil as the magnetic flux through it changes. The magnet is accelerating under gravity: it leaves the coil slightly faster than it enters. As Figure 20.2 shows, the maximum induced e.m.f. is greater as the magnet exits, although this e.m.f. is induced for a shorter time. If the magnet were dropped from a greater height, the e.m.f.s induced would be greater, because the rates of change of flux on entering and leaving would be greater. But the e.m.f.s would last for a shorter time, because the magnet would be travelling faster.

If the terminals of the coil are connected together, instead of to the data-logger, an induced current flows through the coil when the magnet is dropped. The top of the coil becomes a south pole as the south pole of the magnet approaches (its entry is opposed). The bottom of the coil becomes a south pole as the magnet's north pole recedes (its exit is opposed).

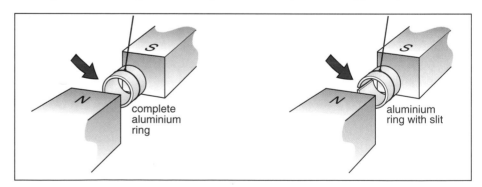

Figure 20.3 Induced currents in the complete ring oppose its motion

Figure 20.3 shows a situation where induced currents have very marked effects. Two aluminium rings are swinging between the poles of a magnet. The ring with a slit keeps swinging for a long time. The complete ring stops quickly. In both rings, a voltage is induced when they swing through the magnetic field. But in the complete ring a current can flow, which opposes the change producing it – the swinging of the ring through the field.

Cutting magnetic fields

Figure 20.4 shows a conductor of length l moving at speed v at 90° to a magnetic field B. In 1 s, the conductor moves a distance v, passing through an area lv of magnetic field. We have

$$\text{induced e.m.f.} = -N \, d\Phi/dt$$

Since $\Phi = BA$ and the magnetic field is constant, this becomes

$$\text{e.m.f.} = -NB \, dA/dt$$

Here there is a single piece of wire, so $N = 1$. Also $dA/dt = lv$. So

$$\text{e.m.f.} = -Blv$$

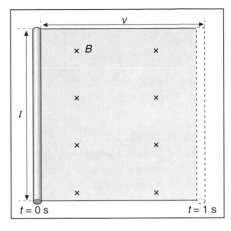

Figure 20.4 For conductor of length l moving at speed v, dA/dt = lv

The Earth's magnetic field

The Earth's magnetic field slopes, so has a horizontal and a vertical component. If you move a wire vertically, it cuts the horizontal component of the Earth's magnetic field. If you move a wire horizontally, it cuts the vertical component of the Earth's field.

Cutting the Earth's magnetic field

- This outdoor experiment requires you to work in a large group. Arrange 100 m of insulated wire into a square of side 25 m. Connect a sensitive voltmeter to the mid-point of the northern side (Figure 20.5).
- Hold the length AB, and quickly jump up with it from ground level to full stretch. Record the time taken to raise the wire and the reading on the voltmeter as you do so. Estimate the distance through which the wire was raised. Calculate its average speed.
- Hold the length AB at a constant height and run with it towards the meter. Record the voltmeter reading and the time taken to cover the distance of 25 m. Calculate the average speed.
- Calculate the horizontal and vertical components of the Earth's magnetic field. Combine these to obtain the Earth's resultant magnetic field.

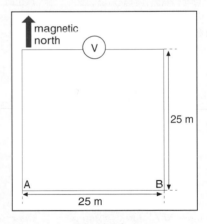

Figure 20.5 Digital voltmeter measures induced e.m.f.

Angle of dip

Figure 20.6 shows the horizontal and vertical components of the Earth's magnetic field in Britain. Their resultant is at an angle of about 70° to the horizontal. This is known as the angle of dip.

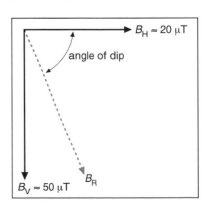

Figure 20.6 In Britain, the angle of dip is about 70°

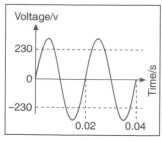

Figure 20.7 This alternating voltage is constantly changing direction at a frequency of 50 Hz

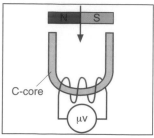

Figure 20.8 Here is a C-shaped piece of iron with a coil of wire wrapped around it. The piece of iron is called a C-core. If you move the magnet down, the field through the iron increases and an e.m.f. is induced

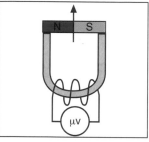

Figure 20.9 If you move the magnet up, the field also changes, which induces an e.m.f. in the opposite direction

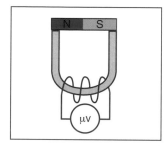

Figure 20.10 If you leave the magnet on the C-core, there is a constant field through it, and no e.m.f. is induced

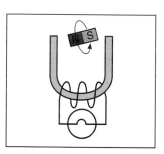

Figure 20.11 You can spin a magnet between the C-cores, constantly changing the field one way and then the other. This is how a generator works

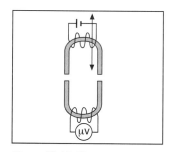

Figure 20.12 You can move an electromagnet up and down. This will produce a changing field and induce a changing e.m.f.

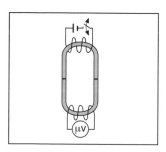

Figure 20.13 Instead of moving the electromagnet up and down, you can just switch it on and off. This will produce a changing field, and therefore induce a changing e.m.f.

Alternating voltage and current

The voltage of a battery is in one direction only. It pushes charge in this direction. This produces a **direct current**, a current in one direction only.

The voltage from the mains supply is **alternating**. This means that it is constantly changing direction (Figure 20.7). It pushes charge first one way and then the other. It produces an **alternating current** (a.c.).

A direct current can produce a steady magnetic field. An alternating current will produce a changing magnetic field. This will, of course, induce an e.m.f in any circuit through which the alternating field passes. All mains supplies nowadays use a.c. because the induced e.m.f's are very useful.

The transformer

Study Figures 20.8 to 20.14 which explain the action of a transformer. The output voltage of a transformer depends on the input voltage. It also depends on the number of turns. The two coils on a transformer are called the primary (input) and the secondary (output). The ratio of the secondary voltage V_s to the primary voltage V_p is the ratio of the number of turns on the secondary N_s to the number of turns on the primary N_p.

$$\frac{V_s}{V_p} = \frac{N_s}{N_p}$$

Transformers are used in many electrical appliances to step the voltage up or down. For instance, if you want to use a 230 V mains supply to power a 12 V train set,

$$\frac{V_s}{V_p} = \frac{12\ V}{230\ V} = 0.052$$

If the primary has 1840 turns, the secondary will need 96 turns.

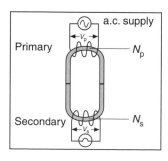

Figure 20.14 An a.c. supply produces a constantly changing field. This induces a constantly changing e.m.f. This is a transformer

Accelerators

The only way to find how matter is constructed, is to break it apart with fast-moving particles produced by an accelerator.

The Fermilab Linear Accelerator (or Linac).

Electrons

Television sets, computer monitors and oscilloscopes all use beams of electrons. However, fast-moving electrons have more exotic uses – to aid the particle physicist in exploring the structure of matter.

The simple **electron gun** is the easiest way to make electrons move quickly.

An electric field will push positive charges along the field lines from positive to negative (Figure 21.1). As the field pushes the charge, potential energy decreases. If the charge is unrestrained, its kinetic energy increases by the same amount. If the charge is negative, the same thing happens except that the charge moves in the opposite direction from negative to positive.

In the electron gun shown in Figure 21.2, the low-voltage supply powers the heater, which heats the cathode. When the cathode gets hot, its atoms vibrate more vigorously, throwing off some of the electrons. This is called **thermionic emission**. The high-voltage power supply attracts electrons from the cathode to the anode. Some pass through the hole in the anode and produce an electron beam. The beam produces a visible spot when it hits the fluorescent screen.

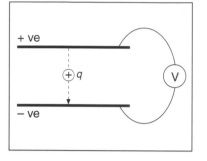

Figure 21.1 When charge moves between the plates, work done = qV

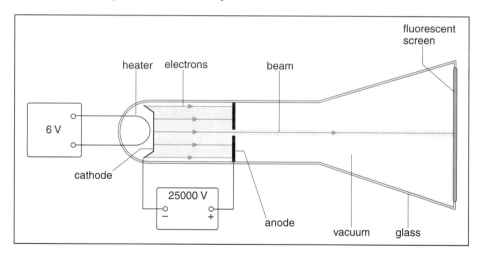

Figure 21.2 Electrons gain kinetic energy as they are attracted from the cathode to the anode

The potential energy lost as the electrons go from the cathode to the anode is equal to the kinetic energy gained:

$$qV = \tfrac{1}{2}mv^2$$

The charge and mass of an electron are $(-)1.6 \times 10^{-19}$ C and 9.1×10^{-31} kg respectively. In a computer monitor, the electrons typically are accelerated through 25 kV. So PE lost = KE gained:

$$qV = \tfrac{1}{2}mv^2 \qquad \text{so} \qquad v^2 = 2qV/m$$

$$\text{and} \quad v = \sqrt{\frac{2qV}{m}}$$

$$= \sqrt{\frac{2 \times 1.6 \times 10^{-19}\text{ C} \times 25\,000\text{ V}}{9.1 \times 10^{-31}\text{ kg}}}$$

$$= 9.3 \times 10^7 \text{ m s}^{-1}$$

If you want to make electrons move faster, you need a higher accelerating voltage. Ordinary power supplies can provide several thousand volts. Higher voltages than this can be provided by a Van de Graaff generator – a bigger version of the demonstration model you may have at school. Figure 21.3 shows a Van de Graaff generator that was used by two British physicists in the 1930s.

Figure 21.3 A Van de Graaff generater powered this 1932 accelerator.

Increasing mass

The simple equation for calculating the speed of an electron does not work when electrons are moving quickly. Consider an electron accelerated through 300 000 V.

$$v^2 = \sqrt{\frac{2qV}{m}}$$

$$= \sqrt{\frac{2 \times 1.6 \times 10^{-19}\text{ C} \times 300\,000\text{ V}}{9.1 \times 10^{-31}\text{ kg}}}$$

$$= 3.25 \times 10^8 \text{ m s}^{-1}$$

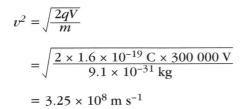

Figure 21.4 The Stanford linear accelerator is 3 km long and can accelerate electrons to 50 GeV

This predicts that the electron will be travelling faster than the speed of light, which is not possible. As a body approaches the speed of light its mass increases. This makes it harder to accelerate. As the accelerating voltage increases it gives the electron more energy. The equation, kinetic energy $= \tfrac{1}{2}mv^2$, still holds, but there is a significant increase in m, and very little increase in v.

The linac

Particle physicists want higher energy particles to smash atoms and smaller sub-atomic particles. To do this they need particles with higher energy (even though these particles are all travelling near the speed of light). One way of accelerating the particles through larger voltages is to use a linear accelerator or **linac**. This accelerates charged particles along an evacuated tube (Figure 21.4). It has a straight tube and a series of charged plates. The voltage between the plates switches from positive to negative repeatedly. This continually accelerates those particles which travel through the holes from one stage to another when the voltage switches (Figure 21.5).

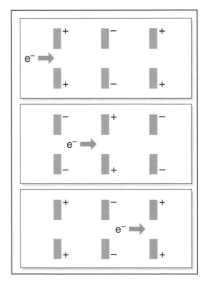

Figure 21.5 The principle of the linear accelerator

Magnetic force on a charged particle

Forces on charged particles

In *Electricity and Thermal Physics* you learnt that a current consists of moving charged particles. Chapter 12 derived the equation ($I = nAqv$) for a current I in terms of the number of moving charged particles per unit volume n, their drift speed v, the area of cross-section of the conductor A and the charge of each particle q.

You know that the force acting on a current flowing at 90° to a magnetic field is given by $F = BIl$. We can substitute for I, so in terms of the moving charged particles

$$F = B(nAqv)l = nAlBqv$$

But Al is the volume of the conductor in the magnetic field, and nAl is the total number of moving charged particles in that volume. Therefore

$$\text{force on } nAl \text{ moving charged particles} = (nAl)Bqv$$

so $\qquad\qquad\qquad$ force on each particle $= Bqv$

This force is at right angles to the direction of motion of the particle.

Electron beams in magnetic fields

- Set up the fine beam tube to project a beam of electrons vertically upwards as shown in Figure 22.1
- Pass a steady current through a pair of coils to produce a uniform horizontal magnetic field across the tube.
- Observe the deflection of the electron beam as it passes through the magnetic field. Reverse the current through the coils and observe the effect.
- Investigate the effect of the magnetic field strength and the accelerating voltage on the electrons' path.

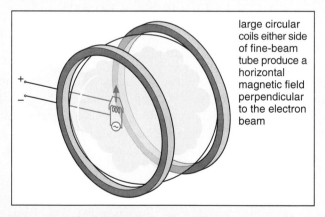

large circular coils either side of fine-beam tube produce a horizontal magnetic field perpendicular to the electron beam

Figure 22.1 The vertical electron beam leaves the gun and enters the horizontal magnetic field between the two coils

Circular paths

You read in Chapter 21 about electron guns. A fine beam tube consists of an electron gun within a glass tube. The tube also contains a gas (often hydrogen) at low pressure. Inelastic collisions between beam electrons and gas atoms result in photons of light being emitted, which show up the electrons' path.

The electron beam in Figure 22.2 is travelling perpendicularly to a magnetic field. This produces a force on the electrons that is perpendicular to their direction of motion, as shown in Figure 22.3. This causes the electrons to perform circular motion.

The magnetic field provides the centripetal force needed for this circular motion. Since this force acts at 90° to the direction of motion, it does no work on the electrons. Their speed remains constant while their velocity changes.

The force acting on each electron, Bqv, is the centripetal force, also equal to mv^2/r:

$$Bqv = mv^2/r \qquad \text{so} \qquad r = mv/Bq$$

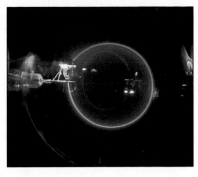

Figure 22.2 The electrons follow a circular path

The cyclotron

The more you want to accelerate a particle with a linear accelerator, the longer the accelerator needs to be. This is inconvenient. It is easier to have the particles go round in a circular path. If they are accelerated during every rotation, they can be given more and more energy without the accelerator needing to be any larger.

Figure 22.4 shows a diagram of a **cyclotron**. It uses a magnetic field to make the particle travel in a circular path within two large D-shaped semi-circular chambers. The D's are connected to a high frequency alternating voltage which alternately pulls and pushes the particles from one D to another.

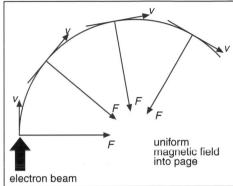

Figure 22.3 The force is always perpendicular to the electron beam

The centripetal force is provided by the magnetic field B through which the particle moves. If the particle has a mass m, a charge q and moves with speed v with a path of radius r,

$$\frac{mv^2}{r} = Bqv$$

$$\therefore \frac{v}{r} = \frac{Bq}{m}$$

v/r is the angular speed ω. The frequency of rotation $f = \omega/2\pi$

$$\therefore f = \frac{\omega}{2\pi} = \frac{Bq}{2\pi m}$$

The frequency of rotation does not depend on the speed of the particle. It depends on B, q and m, all of which are fixed for particles travelling significantly below the speed of light.

The high-frequency accelerating voltage will continually accelerate all particles, whatever their speeds, and they spiral out, their radius increasing, as they accelerate.

Ernest Lawrence invented the cyclotron in 1929. His machine was capable of accelerating particles to more than 1 MeV. Cyclotrons are simple devices and still used in hospitals to produce beams of particles to treat cancer.

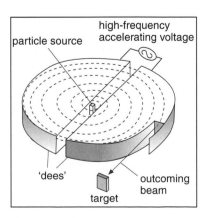

Figure 22.4 The rotation frequency of the particles stays the same. As they get faster, they just spiral out.

Smashing atoms

Figure 23.1 The large ring at CERN in Geneva has a circumference of 27 km and has more than 4000 magnets to control and steer a beam of particles

Figure 23.2 A beam of neutrinos (invisible in picture) entered this bubble chamber and one collision produced this dramatic spray of particles

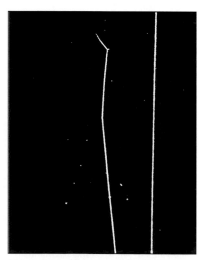

Figure 23.3 Alpha particle tracks in a cloud chamber

You know from *Mechanics and Radioactivity* that physicists use fast moving particles to probe inside matter. They can use beams of electrons, but protons have much higher mass and are more often used. The protons can be accelerated by a linac or cyclotron, but they can be given even higher energies using a synchrotron. In the synchrotron, a magnetic field keeps a bunch of particles moving in a circular path. An alternating electric field accelerates the particles, as in a cyclotron, but the frequency of the electric field is synchronised to take account of the increasing mass of the particles as they approach the speed of light (Figure 23.1).

Fixed targets

The beam of protons collides with something – the target. The simplest target is a bath of liquid hydrogen. This contains many protons – the hydrogen nuclei. The collisions between the fast moving protons of the beams and the fixed protons in the liquid hydrogen produce showers of elementary particles for the physicist to study.

Colliding beams

It is more revealing when two beams of particles travelling in the opposite direction collide. When they do so, the initial momenta are usually equal and opposite; the products can in principle have zero momentum and therefore zero kinetic energy. So all the available energy can be used to create new particles. Collisions are more likely between a beam and a dense fixed target than between two beams, but in modern ring accelerators the beams can be steered very precisely and can be made to cross and collide very many times each second.

Particle detectors

Charged particles *ionise* the atoms of liquid or gas through which they pass. The number of ions produced depends on charge and velocity. A *bubble chamber* is filled with liquid hydrogen and charged particles passing through leave a trail of ions, around which small bubbles form. Photographs taken from different angles show the precise track of each particle (Figure 23.2).

In the laboratory you may have seen the tracks made by alpha particles in a cloud chamber. Cold solid carbon dioxide keeps the bottom of the chamber cold. A felt ring soaked in alcohol keeps the air in the chamber saturated. Alcohol condenses on the ions produced by an alpha particle travelling through the air and makes the path of the alpha particle visible (Figure 23.3).

Spark chambers (Figure 23.4) and *drift chambers* have plates or wires carrying high potentials, so, when particles ionise the gas in the chamber, a current flows between the plates or wires. A computer can quickly reconstruct the paths of the particles and select events that show particular features.

Deflecting radiations in fields

Alpha and beta radiation both have electric charge; they can be deflected by both electric and magnetic fields. Gamma radiation is uncharged and cannot be so deflected.

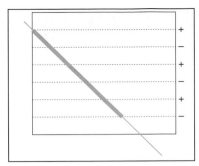

Figure 23.4 A spark chamber

Deflecting radiation with magnetic fields

- Use a source-handling tool to mount a beta source 10 cm from a GM tube and then bring up a magnet as shown in Figure 23.5. Note the change in the count.
- Then move the GM tube into a position like that shown by the dotted lines to find the deflected particles.
- Repeat the experiment with alpha and gamma sources in turn.

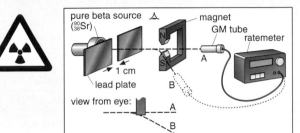

Figure 23.5   Deflecting beta radiation

Figure 23.6 shows the deflections of the radiations in a magnetic field. You need to use Fleming's left-hand rule to check the deflection of the particles. Both alpha and beta-plus are positive. The direction of travel is the direction of the current. They are deflected upwards. Beta-minus radiation is deflected in the opposite direction. Alpha particles are so much more massive than beta particles that their deflection is hardly noticeable; their deflection on the diagram is greatly exaggerated.

Most charged particles have a charge equal or opposite to the electronic charge, so in a known magnetic field the direction and radius of curvature of the tracks give a direct indication of the charge and momentum of the particles (Figure 23.7). Other features allow other properties to be found. A particle with no charge leaves no track but can be detected by a gap or kink in other tracks (Figure 23.8).

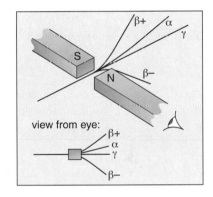

Figure 23.6   Behaviour of radiations in a magnetic field

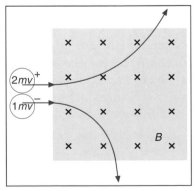

Figure 23.7 The particle with greatest momentum is deflected least

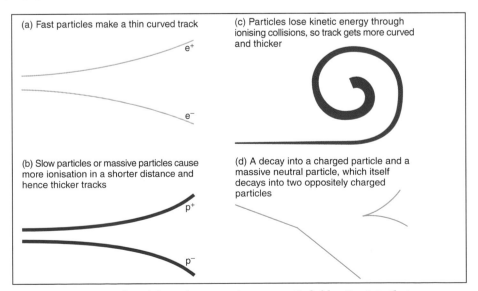

(a) Fast particles make a thin curved track

(b) Slow particles or massive particles cause more ionisation in a shorter distance and hence thicker tracks

(c) Particles lose kinetic energy through ionising collisions, so track gets more curved and thicker

(d) A decay into a charged particle and a massive neutral particle, which itself decays into two oppositely charged particles

Figure 23.8 Features of particle tracks as seen in a magnetic field acting into the page

24 Energy has mass

Which has more mass – a cold potato or the same potato when hot? Well, the hot one has more mass, because the energy it gained getting hot has mass. If a body gains energy, that energy has mass, and the body gains mass as well as energy. You can calculate the increase in mass from the most well-known formula in physics: $E = mc^2$, where E is the energy, c is the speed of light and m is the mass of that energy.

To raise the temperature of 0.1 kg of potato from 20 °C to 100 °C takes about 36 000 J. The increase in mass is given by the formula:

$$m = \frac{E}{c^2} = \frac{36\,000\,\text{J}}{(3 \times 10^8 \text{ m s}^{-1})^2} = 4 \times 10^{-13} \text{ kg}$$

No wonder we do not notice the increase in mass – it is terribly small. But the mass of energy is noticeable in nuclear and particle physics.

Beta decay

You learnt in Chapter 35 of *Mechanics and Radioactivity* that, in beta decay, a neutron in the nucleus splits into a proton and an electron. The mass of the neutron is greater than the mass of the proton plus the mass of the electron. The difference in mass is the mass of the energy released:

neutron	=	proton	+	electron	+	energy
$(1.674\,38 \times 10^{-27} \text{ kg})$	=	$(1.672\,08 \times 10^{-27} \text{ kg})$	+	$(0.000\,91 \times 10^{-27} \text{ kg})$	+	$(0.001\,39 \times 10^{-27} \text{ kg})$

The energy released has a mass of $0.001\,39 \times 10^{-27}$ kg:

$$\text{energy released} = mc^2$$

$$= (0.001\,39 \times 10^{-27} \text{ kg}) \times (3 \times 10^8 \text{ m s}^{-1})^2 = 1.25 \times 10^{-13} \text{ J}$$

The unified mass unit

Accurate measurement of atomic mass is based on the carbon-12 atom, $^{12}_{6}\text{C}$. The unified mass unit, u, is defined as one-twelfth of the mass of a carbon-12 atom. Its value is u $= 1.66 \times 10^{-27}$ kg. The mass of a hydrogen atom is 1.0079 u. Alternatively, you may say that the relative atomic mass of hydrogen is 1.0079. The unified mass unit is a much more convenient unit for calculating mass changes involving sub-atomic particles.

Let us calculate the energy that would have a mass of 1 u.

$$E = mc^2 = 1.66 \times 10^{-27} \text{ kg} \times (3 \times 10^8 \text{ m s}^{-1})^2 = 1.49 \times 10^{-10} \text{ J}$$

in electron volts this is $\dfrac{1.49 \times 10^{-10} \text{ J}}{1.6 \times 10^{-19} \text{ J eV}^{-1}} = 934 \text{ MeV}$

You may have used this value already if you studied either the Astrophysics or the Nuclear and Particle Physics topics in Unit 3.

Nuclear fission

Fission means *splitting up*. Nuclear fission already powers present-day nuclear power stations. In this, large nuclides with mass numbers over 200, usually uranium or plutonium, are bombarded with neutrons to split them into smaller nuclides.

A typical reaction involves the absorption of a neutron by a nucleus of uranium-235:

$$^{235}_{92}\text{U} + {}^1_0\text{n} \rightarrow {}^{141}_{56}\text{Ba} + {}^{92}_{36}\text{Kr} + 3{}^1_0\text{n} + \text{energy released}$$

The absorption of a single neutron leads to three more being released, each of which can cause a further fission reaction. Figure 24.1 shows how this process can lead to a chain reaction.

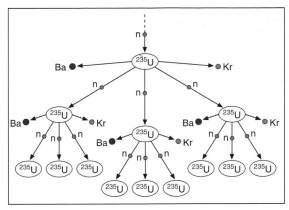

Figure 24.1 Each reaction emits more neutrons, which cause further reactions

Nuclear fusion

Fusion means *joining together*. If you join together light nuclei, large amounts of energy are released. Within the stars and our Sun, hydrogen nuclei join together and release huge amounts of energy.

The fusion process takes place in three stages, as Figure 24.2 shows. The whole fusion reaction can be represented by the equation:

$$4{}^1_1\text{H} \rightarrow {}^4_2\text{He} + 2{}^0_1\beta^+ + 2\nu + \text{energy}$$

Energy is released as gamma photons and kinetic energy of the resulting particles. Table 24.1 gives the masses of the particles involved in the fusion reaction.

How can we apply these numbers to the Sun? When each helium nucleus is formed, four protons, of total mass 4.029 96 u, become a particle of total mass 4.002 55 u. When the positions are taken into account the difference in mass is 0.026 31 u, which is $4.367\,46 \times 10^{-29}$ kg. Using $E = mc^2$, this mass is equivalent to 3.93×10^{-12} J of energy released. Although each fusion reaction releases only this small amount of energy, the very large number of reactions taking place each second in a star results in a very large rate of energy release, e.g. 3.8×10^{26} W for the Sun.

European countries are jointly making large investments in the JET project (Figure 24.3) to research into nuclear fusion as a practicable future power source.

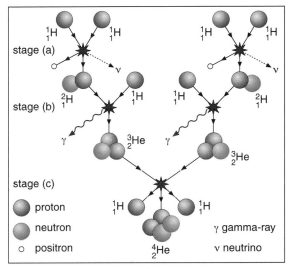

Figure 24.2 Proton–proton chain

Table 24.1 *Masses of particles involved in the proton–proton chain where $1\,u = 1.66 \times 10^{-27}$ kg*

Particle		Mass/u
proton	${}^1_1\text{H}$	1.007 49
helium nucleus	${}^4_2\text{He}$	4.002 55
positron	${}^0_1\beta^+$	0.000 55
neutrino	ν	negligible

Figure 24.3   The Joint European Torus – a test-bed for nuclear fusion

Practice Questions

Chapter 1

1.1 Explain the differences between the terms mass and weight.

1.2 Draw a free-body force diagram for a mass of 5 kg hanging on the end of a spring balance.

1.3 What is a gravitational field? Define gravitational field strength.

1.4 A mass of 20 kg on the surface of the Moon is attracted towards the Moon with a force of 32 N. Calculate the gravitational field strength of the Moon at its surface. What will be the force of attraction between the Moon and a 70 kg astronaut standing on its surface?

1.5 Show that the units of gravitational field strength are equivalent to those of acceleration. Use the data in the chapter to find the acceleration of the Moon due to the attraction between it and the Earth.

Chapter 2

2.1 State Newton's law of gravitation. Show that the unit of the universal gravitational constant is $N\,m^2\,kg^{-2}$. Express this in base units.

2.2 Two identical lead spheres have a combined mass of 12 kg. The density of lead is $11\,400\ kg\,m^{-3}$. Find the radius of each sphere. Calculate the gravitational force of attraction between the two spheres when they are touching.

2.3 Two masses, one of 300 g and one of 500 g, are placed 40 cm apart and a 75 g mass is placed midway between them. Find the magnitude and direction of the resultant gravitational force acting on the 75 g mass produced by the presence of the other masses.

2.4 One method for measuring the universal gravitational constant G involves mounting a simple pendulum from a rigid support close to the side of a mountain. Gravitational attraction between the pendulum bob and the mountain pulls the pendulum away from the vertical as shown in the diagram.

A suitable mountain for this experiment (and a very good climb!) is Schiehallion in Scotland. Draw a free-body force diagram to show the forces acting on the pendulum bob. The volume of Schiehallion is $1.6 \times 10^9\ m^3$ and its average density is $3000\ kg\,m^{-3}$. Find its mass. Calculate the gravitational force it would exert

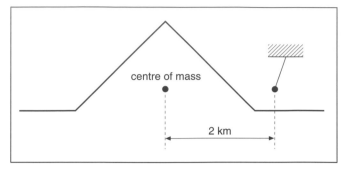

on a 2 kg pendulum bob placed as shown in the diagram. Find the angle from the vertical through which the pendulum would be deflected.

2.5 Question 2.4 assumes that the centres of mass of the bob and the mountain are horizontally aligned. In practice, the mountain's centre of mass is much lower. Discuss the effect that this will have on the angle through which the pendulum is deflected.

Chapter 3

3.1 A satellite is continuously in free fall towards the centre of the Earth. Explain why it remains the same distance above the Earth's surface and why its speed is unaffected by the gravitational force acting on it.

3.2 The Earth's orbital radius around the Sun is 1.5×10^{11} m and 1 Earth year is 3.2×10^7 s. Use the values in Table 3.1 to plot a graph of $(period/s)^2$ against $(radius\ of\ orbit/m)^3$ for the planets. How does your graph show that T^2 is directly proportional to r^3?

3.3 Show that the slope of a graph of $(period\ of\ orbit)^2$ against $(radius\ of\ orbit)^3$ for any group of bodies orbiting the same central mass is constant. Use the gradient of your graph in question 3.2 to determine the mass of the Sun.

3.4 Calculate the period of a near-Earth satellite, i.e. one where $g \approx 9.8\ N\,kg^{-1}$ and orbital radius $r \approx 6400$ km.

3.5 What is a geostationary satellite? Calculate the height of such a satellite above the Earth's surface.

Chapter 4

4.1 The gravitational field strength of the Earth is $0.0028\ N\,kg^{-1}$ at the distance of the Moon's orbit. The mass of the Earth is 6.0×10^{24} kg. Calculate the radius of the Moon's orbit.

4.2 The Earth's gravitational field obeys an inverse square law. What is an inverse square law?

4.3 A satellite orbits the Earth (radius 6.4 Mm) at a height of 9.6 Mm above the Earth's surface. Find the ratio between the radius of the orbit and that of the Earth. The acceleration due to gravity at the Earth's surface is 9.8 m s^{-2}. Calculate the centripetal acceleration of the satellite. Hence find its orbital speed and its period.

4.4 The mass of the Earth is 6.0×10^{24} kg and its radius is 6400 km. Show that the Earth's gravitational field can be considered to be uniform from its surface up to a height of 10 km.

4.5 Calculate the work done against gravity in raising a mass of 500 kg from the Earth's surface to a height of 5 km. How will this value compare with that required to raise the same mass from a height of 50 km to one of 55 km? Explain your answer.

Chapter 5

5.1 A certain coulombmeter consists of a digital voltmeter connected across a capacitance of 0.10 µF. What will be the voltage across the capacitor when the display indicates a charge of 40 nC?

5.2 State Coulomb's law. Express the units of the constant k, associated with Coulomb's law, in terms of base units only.

5.3 Criticise the statement 'an insulated metal sphere is given a positive charge'. The centre of a sphere with a charge of +95 nC is placed 12 cm from the centre of another sphere with a charge of +106 nC. Calculate the electrostatic force between them.

5.4 Compare your answer to question 5.3 with the gravitational force between the spheres, given that they each have a mass of 0.4 kg.

5.5 Two identical, charged spheres are suspended from the same point on two insulating threads as shown in the diagram.

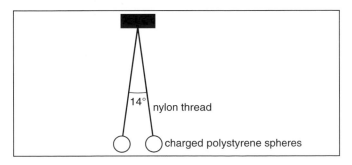

Each sphere has a mass of 500 mg and carries a charge of 50 nC. Draw a free-body force diagram for the sphere

on the left. Calculate the tension in its thread and the size of the electrostatic force of repulsion acting on it. The centres of the spheres are 20 cm apart. Calculate a value for the permittivity of air.

Chapter 6

6.1 What is an electric field? Define electric field strength and state its unit.

6.2 Express the unit of electric field strength in terms of only base units. Explain why a very small test charge must be used when measuring electric field strength.

6.3 Calculate the electric field strength 15 cm from the centre of a charge of 35 nC. What will it be at distances of (a) 30 cm, (b) 45 cm?

6.4 Describe how you would investigate the shape of an electric field. Sketch the electric field pattern associated with (a) a positive point charge, (b) a negative point charge.

6.5 One model of a hydrogen atom describes it as an electron orbiting a central proton. What provides the force required for the circular motion of the electron? What effect does this force have on the speed of the electron? Explain your answer. What happens to the electron's velocity?

Chapter 7

7.1 Sketch the electric field pattern associated with two oppositely charged parallel plates. With reference to your sketch, describe what is meant by a uniform electric field.

7.2 The potential difference across two parallel plates is 180 V. Calculate the energy transferred to a charge of +500 pC when it moves from the positive to the negative plate. If the plates are 9 cm apart, calculate the average force acting on the charge as it moves between the plates.

7.3 What is an equipotential? Describe how you would investigate the shape of the equipotentials between two oppositely charged parallel plates.

7.4 Redraw your field pattern sketch from question 7.1 and add equipotentials to it. How are the directions of the field lines and the equipotentials related?

7.5 Two long, parallel metal plates are 5 cm apart and have a potential difference of 1 kV across them. Calculate the electric field strength between the plates. An electron (charge -1.6×10^{-19} C and mass 9.1×10^{-31} kg) is placed centrally between the plates. Calculate the force acting on the electron and its initial acceleration. Explain how the acceleration varies as the electron moves away from its central position.

PRACTICE QUESTIONS

Chapter 8

8.1 Redraw your field pattern sketch from question 6.4(a) and add equipotentials to it. What is the shape of these equipotentials? Why do they get further apart as the distance from the charge increases?

8.2 Point Y has a gravitational potential that is 4.5×10^3 J kg^{-1} greater than point X. Calculate the work done when a mass of 8 kg is moved from X to Y: (a) along a straight line joining the two points, (b) along a winding path that is three times the length of the direct route.

8.3 What is meant by escape speed? About 3 MJ is needed to completely free 1 kg from the Moon. Calculate the approximate value of the escape speed for an object on the Moon.

8.4 State two similarities and two differences in the properties of gravitational and electric fields.

8.5 The electric field between two oppositely charged parallel plates is uniform. Explain the circumstances under which a gravitational field can be considered to be uniform. When a mass m falls a distance h in a uniform gravitational field of strength g, the energy transferred to it is mgh. State a similar equation for the energy transferred to a charge Q when it 'falls' a distance x in a uniform electric field of strength E.

Chapter 9

9.1 Describe the structure of a capacitor.

9.2 A capacitor is connected in series with a battery, a resistor and a micro-ammeter. Explain why a current flows initially in the rest of the circuit despite the fact that no charge is able to pass directly between the capacitor's plates.

9.3 Draw the circuit that you would use to investigate how current varies with time as a capacitor is charged from a battery in series with resistor. State the readings that you would take and describe how you would expect the current to vary.

9.4 Sketch a current–time graph for a capacitor-charging circuit. What two factors determine the initial value of the current? A 500 μF capacitor in series with a 50 kΩ resistor is charged using a 12 V battery. Calculate the initial charging current.

9.5 Explain why a capacitor takes a longer time to charge to the same potential difference when it is in series with a larger resistance.

Chapter 10

10.1 Explain why the total charge on a capacitor is always zero no matter what potential difference is across it.

10.2 The table gives the current in a capacitor circuit at 10 s intervals from when it is first connected.

time/s	0	10	20	30	40	50	60	70	80	90	100	110	120
current/μA	80	65	52	42	34	28	22	18	15	12	10	8	6

Plot a graph of current against time. Use your graph to find an approximate value for the charge on either plate of the capacitor.

10.3 Define capacitance. Express the unit of capacitance in terms of base units.

10.4 The capacitor in question 10.2 has a 50 kΩ resistor connected in series with it. Calculate the potential difference across the capacitor when it becomes fully charged. (Hint: this will be the same as the e.m.f. of the supply). Hence, find the approximate value of the capacitance of the capacitor used in this circuit.

10.5 Describe, with the aid of a circuit diagram, how you would keep the current in a circuit constant as a capacitor is charged. How long will it take to charge a 500 μF capacitor to a potential difference of 12 V using a constant current of 40 μA.

Chapter 11

11.1 Sketch a graph showing how the potential difference across the capacitor in Figure 11.1 on page 24 changes with time as the capacitor is charged. Add a second line to your graph to show how the potential difference across the resistor varies during the same period. Which of your lines has the same shape as that of the current–time graph for the circuit? Explain why they have the same shape.

11.2 Repeat question 11.1 to show how the voltages vary as the capacitor is discharged.

11.3 Sketch, on the same axes, current–time graphs for a high- and a low-value capacitor charged by the same supply through identical resistors.

11.4 The time constant of a resistor-capacitor circuit is 15 s. Explain what is meant by time constant.

11.5 A 250 μF capacitor is charged through a 100 kΩ resistor. Calculate the time constant of the circuit. The initial current is 90 μA, what will it be after 50 s? What will be the final voltage across the capacitor?

Chapter 12

12.1 A 2200 µF capacitor is charged to a potential difference of 5.0 V. Calculate the charge on one of its plates at the end of the charging process. The capacitor is then discharged through a 15 kΩ resistor. How much of the charge will remain after 20 s?

12.2 The decay constant of a radioactive isotope of strontium is 7.8×10^{-10} s^{-1}. A source initially contains 4.5 µg of this strontium isotope. Calculate the mass of this isotope remaining after 14 years.

12.3 If a decay process starts with N_0 atoms, how many atoms of that nuclide will be present when $t = t_{\frac{1}{2}}$ (give your answer in terms of N_0)? Substitute these 'values' for N and t into the equation $N = N_0\,e^{-\lambda t}$ and derive the relationship $\lambda t_{\frac{1}{2}} = \ln 2$.

12.4 A student obtains the following results from an experiment involving the decay of a radioactive source.

time/s	0	20	40	60	80	100	120	140	160	180	200
activity/Bq	820	530	330	210	145	95	59	35	22	14	9

Plot a graph of ln(activity/Bq) against time. Use your graph to find the decay constant and, hence, the half-life of this radioactive source.

12.5 Calculate the half-life of the capacitor discharge in question 12.1.

Chapter 13

13.1 A spring extends by 4.5 cm when supporting a load of 18 N and by 9.0 cm when supporting a load of 36 N. Calculate its spring constant. How much energy is stored in this spring when the force on it is 25 N?

13.2 Describe how you would show experimentally that the energy stored in a charged capacitor is directly proportional to the square of the potential difference across it.

13.3 When a capacitor is charged to a potential difference of 50 V, the magnitude of the charge on each of its plates is 10 µC. Calculate its capacitance and the energy stored.

13.4 A mass is placed on the end of a vertical spring and gently lowered to its equilibrium position. Compare the energy stored in the stretched spring with the gravitational potential energy lost by the mass and account for any difference.

13.5 Explain whether a large value capacitor is analogous to a strong or a weak spring.

Chapter 14

14.1 Calculate the combined capacitance when a 150 µF capacitor is connected in parallel with a 350 µF capacitor. What would be the combined capacitance if they were connected in series?

14.2 You are provided with three 500 µF capacitors which can be used individually, in pairs or in combinations involving all three. Sketch each of the possible capacitor combinations, calculating the combined capacitance in each case.

14.3 A 330 µF capacitor is connected in parallel to a 470 µF capacitor and a 12 V supply. State the potential difference across each capacitor. Calculate the charge on one plate of the 330 µF capacitor and the energy stored in the 470 µF capacitor.

14.4 A 50 µF capacitor is connected in series with a 150 µF capacitor and a 9 V battery. Calculate their combined capacitance. For each capacitor, find the charge on one plate, the potential difference and the energy stored.

14.5 A 15 µF capacitor is charged to a potential difference of 50 V and disconnected from the supply. Calculate the charge on one plate and the energy stored. It is then connected across (in parallel to) an uncharged 25 µF capacitor. The charge redistributes between the two capacitors but the total charge remains constant. Calculate (a) the combined capacitance of the two capacitors, (b) the final potential difference across them, (c) the final charge on each capacitor, (d) the total energy stored by the two capacitors. Comment on the difference between the energy stored before and after connection to the second capacitor.

Chapter 15

15.1 What is a magnetic field? Describe two methods for finding the shape of a magnetic field. How is the direction of a magnetic field defined?

15.2 Sketch the magnetic field pattern of a single bar magnet. Mark the north and south ends, and indicate the directions of all magnetic field lines.

15.3 Sketch the magnetic field pattern produced by two attracting magnets. Mark the north and south ends, and indicate the directions of all magnetic field lines.

15.4 What is a neutral point?

15.5 The diagram shows a magnet and the direction of the Earth's magnetic field relative to it.

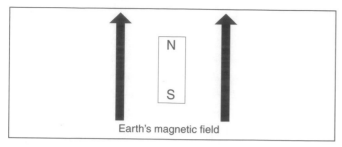

Earth's magnetic field

Make a copy of the diagram and mark on it the positions of any neutral points. Repeat for the situation where the magnet has its south end at the top.

Chapter 16

16.1 Sketch the magnetic field pattern of a long straight current-carrying wire, indicating the directions of both the current and the magnetic field lines.

16.2 What is a solenoid? Sketch the magnetic field pattern of a current-carrying solenoid, indicating the directions of both the current and the magnetic field lines.

16.3 Compare the magnetic field pattern of a solenoid with that of a bar magnet. Describe how a solenoid can be used to produce a bar magnet.

16.4 Determine the polarity of the left-hand end of the current-carrying coil in Figure 16.3 on page 37. Explain how you arrived at your answer.

16.5 A second coil is placed close to the one shown in Figure 16.3 on page 37, such that their central axes align. Explain what happens if the current in the second coil is in the same direction as that in the original coil. What would be the effect of reversing the current in (a) one of the coils, (b) both coils.

Chapter 17

17.1 State the similarities and the differences between a bar magnet and a magnadur magnet. Describe how you would produce a uniform magnetic field.

17.2 With the aid of a diagram, explain how a force arises from the combination of a uniform magnetic field and the circular magnetic field of a current-carrying wire.

17.3 Copy Figure 17.3 on page 38 and add to it the positions of any neutral points.

17.4 The diagrams show three situations involving a wire carrying a current through a magnetic field. For each situation find the direction of the stated quantity.

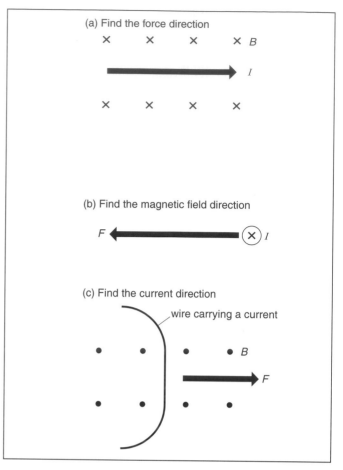

(a) Find the force direction

(b) Find the magnetic field direction

(c) Find the current direction

wire carrying a current

17.5 Describe the experiment that you would use to show that the force exerted on a conductor carrying a current in a magnetic field is proportional to the current flowing. What other factors does this force depend on?

Chapter 18

18.1 Define the tesla. Express the tesla in base units.

18.2 The speech coil in a small loudspeaker has a diameter of 4 cm and consists of 300 turns. Calculate its total length. The speaker magnet produces a magnetic field of strength 200 mT that, by being radial, is always perpendicular to the total length of the coil. Calculate the force on the speech coil when it carries a current of 5 mA. Explain what happens to the speech coil when an alternating current flows in it.

18.3 A rigid wire loop is connected in series to a 6.0 V battery and a 6.0 V, 9.0 W lamp. Calculate the size of the current. The circuit stands on a top-pan electronic balance as shown.

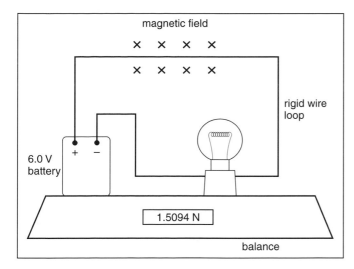

With reference to the diagram, determine the direction in which the current flows through the top part of the wire. A uniform horizontal magnetic field of strength 50 mT acts at right angles to 100 mm of the top part of the wire, as shown. Calculate the magnitude of the force acting on the wire and state its direction. The reading on the balance is 1.5094 N. What will this reading become if the direction of the magnetic field is reversed?

18.4 A solenoid is 35 cm long and has 100 turns. Calculate the magnetic field strength (magnetic flux density) inside it when a current of 4.2 A flows.

18.5 The magnetic flux density produced by the current flowing in a long straight wire is 6 μT at a distance of 15 cm from it. Calculate the size of the current. State one other effect that this flow of current will have on the wire.

Chapter 19

19.1 This question concerns Figure 19.1c on page 42. State and explain fully what happens when the switch (a) is first closed, (b) has been closed for several seconds, (c) is opened.

19.2 In Britain, the Earth's magnetic field dips downwards towards the north at about 70° to the horizontal. As explained in Chapter 20, this magnetic field is often considered as two components, a vertical component of strength 50 μT and a horizontal component of strength 20 μT. A laboratory is 9.2 m long by 7.5 m wide and has a height of 2.4 m. Its longest side lies in a north–south direction. Calculate both the vertical and horizontal components of the magnetic flux through this laboratory as a result of the Earth's magnetic field.

19.3 State both Faraday's and Lenz's laws of electromagnetic induction. What conservation law is Lenz's law a consequence of?

19.4 An isolated flat coil of wire has an area of cross-section of 4.0 cm². The coil has 200 turns. It is placed with a magnetic field of strength 450 mT passing directly through it. Calculate the magnetic flux through a single turn and the total magnetic flux linking the coil. The magnetic field through the coil is reduced to zero in 0.25 s. Calculate the average rate of reduction of the total magnetic flux linking the coil during this process. State the e.m.f. induced across the ends of the coil by this process.

19.5 A coil of wire has 80 turns, each of area 40 cm². It is placed with a magnetic field of flux density 150 mT passing directly through it. Calculate the e.m.f. induced across the ends of the coil when the magnetic field (a) falls to zero in a time of 50 ms, (b) increases to a strength of 240 mT in a time of 150 ms, (c) is reversed in a time of 300 ms. Explain which of the induced e.m.f.s will have the opposite polarity to the other two.

Chapter 20

20.1 The diagram shows a gold ring on a silk thread about to swing through a magnetic field.

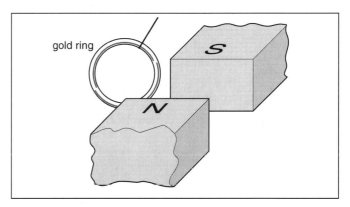

Two students argue about the influence that the magnetic field will have on the motion of the ring. Student A says that the effect of the field is to increase the speed of the ring as it enters the field and to slow it down as it leaves so that the net effect is zero. Student B says the only effect is that of slowing the ring down. Which student is correct? Explain why the ring behaves as this student describes.

20.2 Some vehicles use an electromagnetic braking system to help them slow down. A metal disc, attached to the axle, rotates near a large electromagnet. Explain why turning the electromagnet on slows the vehicle. Why can such a system not be used as the parking brake for the vehicle?

PRACTICE QUESTIONS

20.3 An eager student cycles to college at 12 m s^{-1}. The metal handlebars are 60 cm wide and cut through the vertical component of the Earth's magnetic field that has a flux density of 50 μT. Calculate the e.m.f. induced across the handlebars. Explain why no current would flow through the student when he touched the metal parts of the handlebars even if he were a good electrical conductor.

20.4 Explain the difference between a direct and an alternating current. State whether the current represented by each of the four graphs is direct or alternating.

Calculate the frequency of any alternating current.

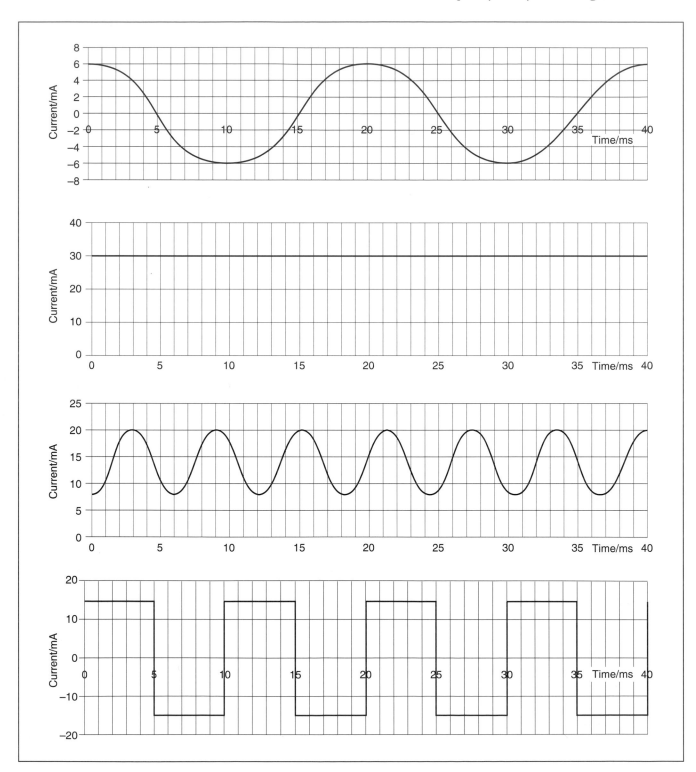

20.5 Describe the basic structure of a transformer and explain its action. A transformer has a 140-turn primary coil and a 700-turn secondary. The primary coil is connected to a 2.5 V a.c. supply. Calculate the output voltage from the secondary coil.

Chapter 21

21.1 An electron gun is used to make a beam of fast-moving electrons. Explain how the electrons are produced and then accelerated.

21.2 Two parallel metal plates have a potential difference of 400 V across them. A particle with a charge of $+4.8 \times 10^{-19}$ C is released next to the positive plate. Explain what happens to this particle. Calculate the maximum kinetic energy the particle can gain and state its position when it has this maximum kinetic energy.

21.3 An ion in a vacuum is accelerated from rest through a potential difference of 350 V, as a result of which it gains 700 eV of kinetic energy. What is the magnitude of the charge on the ion? The mass of the ion is 6.6×10^{-27} kg. Calculate its final speed.

21.4 A proton (charge $+1.6 \times 10^{-19}$ C and mass 1.7×10^{-27} kg) is accelerated from rest in a vacuum through a potential difference of 300 V. How much energy (both in electronvolts and joules) does the proton gain? Calculate the final speed of the proton. How would the energy gained by this proton and its final speed compare with those of an alpha particle (twice the charge and approximately four times the mass) when accelerated under identical conditions?

21.5 Explain why, irrespective of how large a potential difference is used, there is an upper limit to the speed that can be achieved by an accelerating particle and state its value. Explain how a linear accelerator produces charged particles moving at speeds close to this limit.

Chapter 22

22.1 Show how the expression for the magnetic force on a charged particle (Bqv) is derived from that for the magnetic force on a current-carrying wire (BIl).

22.2 An alpha particle has a charge of 3.2×10^{-19} C. Calculate the force that acts on it when it moves at a speed of 1.8×10^7 m s^{-1} at 90° to a magnetic field of strength 120 mT. State the direction of this force.

22.3 Explain why a charged particle moving at 90° to a magnetic field follows a circular path. Suggest what shape of path is followed when moving at 70° to the field.

22.4 In a fine beam tube, the electrons are accelerated through a potential difference of 250 V. The electrons then travel through a magnetic field that has a flux density of 0.92 mT. The electronic charge and mass of an electron are 1.6×10^{-19} C and 9.1×10^{-31} kg respectively. Calculate (a) the speed of the electrons leaving the electron gun and (b) the radius of the path they follow in the magnetic field.

22.5 A linac and a cyclotron are both machines used to accelerate charged particles. In what ways do they differ? What is the main advantage of a cyclotron over a linac for accelerating particles to high energies? What is the main disadvantage? Show that the radius of the path followed by a particle in either 'dee' of a cyclotron is directly proportional to the speed at which it enters that 'dee'. Compare the times spent in a 'dee' by particles entering it at speeds v and $3v$. What bearing does your comparison have on the frequency of the accelerating voltage across the two 'dees'?

Chapter 23

23.1 How does the operation of a synchrotron differ from that of a cyclotron?

23.2 Explain why heavier particles are more likely to be created when two beams travelling in opposite directions collide than when a single beam collides with a stationary target.

23.3 A cloud chamber is one type of particle detector. Name four others. In what way are all particle detectors similar to each other? Compare the tracks produced by both alpha and beta particles passing through a cloud chamber.

23.4 A beam, consisting of alpha, beta-minus and gamma radiations, enters a uniform magnetic field that is at an angle of 90° to the beam. Sketch the path followed by each of the three radiations. Explain the shape of each path. How would the shapes and the directions of these paths differ if it were a uniform electric field?

23.5 The path followed by a charged particle produced within a bubble chamber is often an inward spiral. Account for the curvature of the path and explain why the track spirals inwards.

Chapter 24

24.1 The specific heat capacity of aluminium is 880 J kg^{-1} K^{-1}. Calculate the energy needed to raise the

PRACTICE QUESTIONS

temperature of a 400 g aluminium block from 15 °C to 80 °C. How much mass will the block gain as a result of this process?

24.2 A polonium atom with a mass of $3.485\ 72 \times 10^{-25}$ kg decays by emitting an alpha particle to a lead atom with a mass of $3.419\ 18 \times 10^{-25}$ kg. The mass of an alpha particle is $6.644\ 32 \times 10^{-27}$ kg. Calculate the mass of the energy released during this decay. How many joules is this equivalent to?

24.3 What is the unified mass unit? Express the masses of the polonium atom, the lead atom and the alpha particle in the previous question in terms of atomic mass units.

24.4 Compare and contrast the processes of nuclear fission and nuclear fusion.

24.5 The table shows the masses, in unified mass units, of the particles involved in the nuclear fission of uranium-235 represented by the equation on page 55.

Particle	Mass/u
Uranium-235	235.04
Barium-141	140.91
Krypton-92	91.91
Neutron	1.01

Calculate the energy, both in joules and electronvolts, released by this fission reaction. Your answer should show that the amount of energy released is very small. How is it therefore that nuclear fission can be used to produce large amounts of energy? Describe some of the precautions that need to be taken when producing energy from nuclear fission.

Assessment questions

The questions covering Unit PHY5 'Fields and Forces' have been chosen or modified to be similar in style and format to those which will be set for the A2 assessment tests. These questions meet the requirements of the assessment objectives of the specification. The questions covering the 'Synthesis' material will help in your revision for question 2 of the PHY6 Unit Test.

Fields and Forces

1 Ganymede is one of the moons of Jupiter. Ganymede has a mass M_G and orbits Jupiter, mass M_J. The radius of the orbit is r.
Write down Newton's law of gravitation as applied to the Jupiter–Ganymede system. **[2]**

Write down an expression for the centripetal force F required to cause Ganymede to orbit Jupiter with an angular speed ω. Show that $r^3\omega^2 = GM_J$ where G is the universal gravitational constant. **[3]**

Ganymede orbits Jupiter once every 7.16 days and the radius of its orbit is 1.07×10^9 m. Calculate the mass of Jupiter. **[3]**

(Total 8 marks)
Edexcel GCE Physics Module Test PH4, June 1999

2 The escape speed v from the surface of a planet can be calculated from $v = \sqrt{(2gr)}$ where g is the acceleration of free fall at the planet's surface and r is the planet's radius. For Earth, the escape speed v is 11 km s^{-1}. Calculate the escape speed for a planet of the same mass as the Earth but twice its radius. **[4]**

The escape speed is independent of the mass of the object being launched. Explain why it is nevertheless desirable to keep the mass of a space probe as small as possible. **[1]**

(Total 5 marks)
Edexcel GCE Physics Module Test PH4, January 2000

3 Two identical table tennis balls, A and B, each of mass 1.5 g, are attached to non-conducting threads. The balls are charged to the same positive value. When the threads are fastened to a point P, the balls hang as shown in the diagram. The distance from P to the centre of A or B is 10.0 cm.

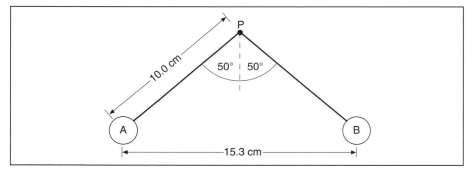

Draw a free-body force diagram for ball A. **[3]**

Calculate the tension in one of the threads. **[3]**

Show that the electrostatic force between the two balls is 1.8×10^{-2} N. **[1]**

Calculate the charge on each ball. **[3]**

State two differences between the electrostatic and the gravitational forces that the two balls exert on each other. **[2]**

(Total 12 marks)
Edexcel GCE Physics Module Test PH4, June 1997

4 A speck of dust has a mass of 1.0×10^{-18} kg and carries a charge equal to that of one electron. Near to the Earth's surface it experiences a uniform downward electric field of strength 100 N C^{-1} and a uniform gravitational field of strength 9.8 N kg^{-1}.

Draw a free-body force diagram for the speck of dust. **[2]**

Calculate the magnitude and direction of the resultant force on the speck of dust. **[4]**

(Total 6 marks)
Edexcel GCE Physics Module Test PH4, January 1997

5 Two parallel plates have a potential difference of 200 V across them. Draw a diagram to show the shape and direction of the electric field between the plates. **[3]**

Add, to your diagram, equipotentials at 50 V and 100 V. **[2]**

(Total 5 marks)
Edexcel GCE Physics Paper 2, June 1994

6 Define the term capacitance. **[2]**

The sockets of modern telephones have six pins. A power supply of 50 V in series with a resistance of about 1000 Ω is connected to pins 2 and 5 as shown.

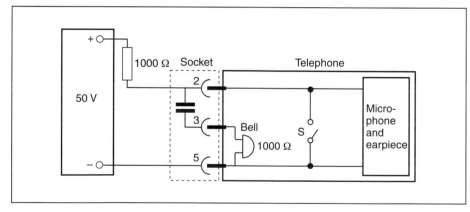

A capacitor of 2 μF is connected between pins 2 and 3. In one installation, a bell of resistance 1000 Ω is connected to pins 3 and 5. Explain why there is a pulse of current through the bell when the circuit is first connected, but not after the bell has been connected for some time. **[2]**

On the circuit, label the values of the voltages across the capacitor and the bell when the circuit has been connected for some time. **[2]**

To dial a number, e.g. 7, the switch S must be closed that number of times. Explain why the bell sounds softly (tinkles) when the switch is closed and opened again. **[2]**

To avoid this tinkling, an 'anti-tinkling' switch is connected to short-circuit the bell during dialling. Add this switch to the circuit. **[1]**

Explain how the anti-tinkling switch stops the bell from tinkling. **[1]**

(Total 10 marks)
Edexcel GCE Physics Module Test PH1, January 1999

7   Most d.c. power supplies include a smoothing capacitor to minimise the variation in the output voltage by storing charge. In a particular power supply, a capacitor of 40 000 μF is used. It charges up quickly to 12.0 V, then discharges to 10.5 V over the next 10 ms, and then charges again to 12.0 V. The process then repeats continually. Calculate the charge on the capacitor at the beginning and at the end of the 10.0 ms discharge period. **[3]**

What is the average current during the discharge? **[3]**

(Total 6 marks)
Edexcel GCE Physics Module Test PH1, January 2000

8   Derive a formula for the equivalent capacitance of two capacitors in series. **[4]**

A 200 μF capacitor is connected in series with a 1000 μF capacitor and a battery of e.m.f. 9 V.

Calculate: (a) the total capacitance **[2]**

             (b) the charge that flows from the battery **[2]**

             (c) the final potential difference across each capacitor. **[3]**

(Total 11 marks)
Edexcel GCE Physics Module Test PH1, January 1996

9   In the circuit shown, switch A is initially closed and switch B is open.

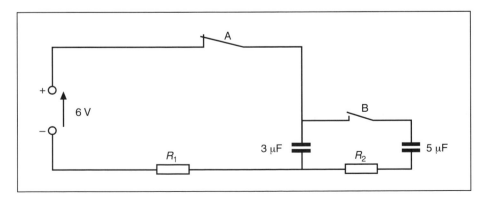

Calculate the energy stored in the 3 µF capacitor when it is fully charged.

[3]

Switch A is now opened and switch B is closed. Calculate the final value of the total energy stored in the two capacitors when the 5 µF capacitor is fully charged.

[4]

State briefly how you would account for the decrease in the stored energy.

[1]

(Total 8 marks)

Edexcel GCE Physics Module Test PH1, June 1999

10 Explain what is meant by a neutral point in a field.

[2]

The diagram shows two similar solenoids A and B.

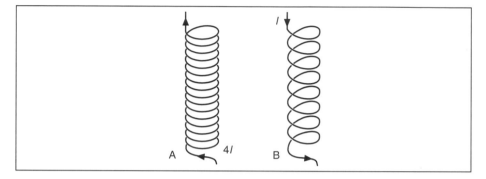

Solenoid A has twice the number of tuns per metre. Solenoid A carries four times the current as B. Draw the magnetic field lines in, around and between the two solenoids.

[4]

If the distance between the centres of A and B is 1 m, estimate the position of the neutral point. Ignore the effect of the Earth's magnetic field.

[3]

(Total 9 marks)

Edexcel GCE Physics Module Test PH4, January 1998

11 The diagram shows a rectangular coil PQRS which can rotate about an axis which is perpendicular to the magnetic field between two magnetic poles.

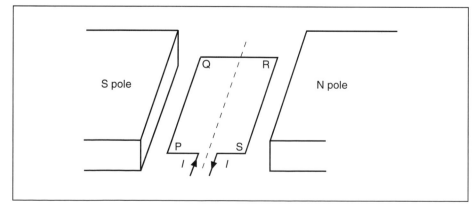

Explain why the coil begins to rotate when the direct current *I* is switched on.

[3]

Add to the diagram an arrow showing the direction of the force on PQ. [1]

State three factors which would affect the magnitude of this force. [3]

A student notices that as the coil rotates faster the current in it reduces. Explain this observation. [2]

(Total 9 marks)
Edexcel GCE Physics Module Test PH4, June 1999

12 A solenoid is formed by winding 250 turns of wire on to a hollow plastic tube of length 0.14 m. Show that when a current of 0.80 A flows in the solenoid the magnetic flux density at its centre is 0.0018 T. [2]

The solenoid has a cross-sectional area of 6.0×10^{-3} m^2. The magnetic flux emerging from one end of the solenoid is 5.4×10^{-6} Wb (*or* 5.4×10^{-6} T m^2). Calculate the magnetic flux density at the end of the solenoid. [2]

Why is the flux density at the end of the solenoid not equal to the flux density at the centre? [1]

(Total 5 marks)
Edexcel GCE Physics Module Test PH4, June 2000

13 A child sleeps at an average distance of 30 cm from household wiring. The mains supply has a nominal voltage of 230 V. Calculate the nominal current when the wire is transmitting 3.45 kW of power. [2]

The maximum value of this alternating current is $\sqrt{2} \times$ its nominal value. Calculate the maximum possible magnetic flux density produced in the region of the child by the current in the cable. [3]

Why might the magnetic field due to the current in the wire pose more of a health risk to the child than the Earth's magnetic field, given that they are of similar magnitudes? [2]

(Total 7 marks)
Edexcel GCE Physics Module Test PH4, January 1997

14 What is meant by the term electromagnetic induction? [3]

Describe an experiment you could perform in a school laboratory to demonstrate Faraday's law of electromagnetic induction. [5]

An aircraft has a wing span of 54 m. It is flying horizontally at 860 km h^{-1} in a region where the vertical component of the Earth's magnetic field is 6.0×10^{-5} T. Calculate the potential difference induced between one wing tip and the other. [2]

What extra information is necessary to establish which wing is positive and which negative? [1]

(Total 11 marks)
Edexcel GCE Physics Module Test PH4, June 1998

15 The graph shows how the magnetic flux density B varies with distance beyond one end of a large bar magnet.

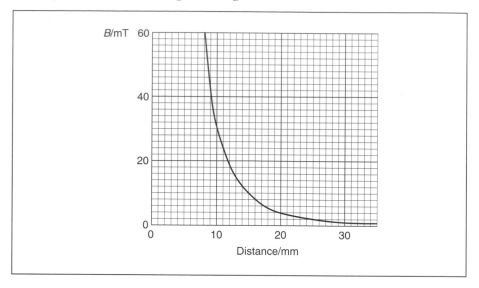

A circular loop of wire of cross-sectional area 16 cm² is placed a few centimetres beyond the end of the bar magnet. The axis of the loop is aligned with the axis of the magnet.

Calculate the total magnetic flux through the loop when it is

(a) 30 mm from the end of the magnet,

(b) 10 mm from the end of the magnet.   [3]

The loop of wire is moved towards the magnet from the 30 mm position to the 10 mm position so that a steady e.m.f. of 15 mV is induced in it. Calculate the average speed of movement of the loop.   [3]

In what way would the speed of the loop have to be changed while moving towards the magnet between these two positions in order to maintain a steady e.m.f.?   [1]

(Total 7 marks)
Edexcel GCE Physics Module Test PH4, June 1999

Synthesis Material

16 Using the usual symbols write down an equation for (a) Newton's law of gravitation, (b) Coulomb's law   [2]

State one difference and one similarity between gravitational and electric fields.   [2]

(Total 4 marks)
Edexcel GCE Physics Module Test PH4, January 1997

17 The equations describing the decay of a sample of a radioactive isotope and the discharge of a capacitor, capacitance C, through a resistor, resistance R, show many similarities. The number of radioactive atoms N

in the sample and the charge Q on the capacitor both decay exponentially. Complete the following table.

	Radioactive decay	Capacitor discharge
Decay law	$N = N_0 e^{-\lambda t}$	$Q = Q_0 e^{-t/RC}$
Rate of decay	$dN/dt = \lambda N$	
Time constant/s		RC
Half-life/s	$\ln 2/\lambda$	
Amount left after two time constants		Q_0/e^2
Amount left after three half-lives	$N_0/8$	

(Total 5 marks)

Edexcel GCE Physics Module Test PH4, June 1997

18 The graph shows the decay of a radioactive nuclide.

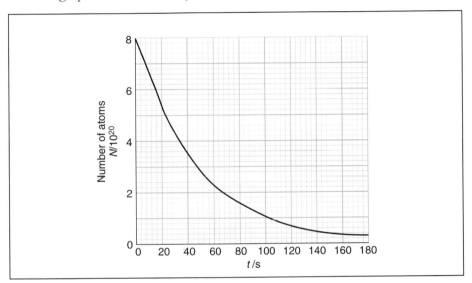

Determine the half-life of this radionuclide. **[2]**

Use your value of half-life to calculate the decay constant λ of this radionuclide. **[1]**

Use the graph to determine the rate of decay dN/dt when $N = 3.0 \times 10^{20}$. **[3]**

Use your value of the rate of decay to calculate the decay constant λ of this radionuclide. **[2]**

Explain which method of determining the decay constant you consider to be more reliable. **[1]**

(Total 9 marks)

Edexcel GCE Physics Module Test PH2, June 2000

19 A 30 μF capacitor is fully charged from a 9 V supply. Calculate the maximum charge stored on the capacitor. **[2]**

The capacitor is discharged through a 20 kΩ resistor. Calculate the time constant for this discharge. **[1]**

How much charge will remain on the capacitor after 1.8 s?   [2]

Sketch a graph to show how the charge stored on the capacitor varies with time during the first 1.8 s of discharge. Label the axes with any significant values.   [3]

(Total 8 marks)
Edexcel GCE Physics Module Test PH4, January 1997

20 An example of a thermonuclear fusion reaction is given by the following equation.

$$^2_1\text{D} + ^3_1\text{T} \rightarrow ^4_2\text{He} + ^1_0\text{n} + \Delta E$$

data: mass of ^2_1D = 2.0136 u

mass of ^3_1T = 3.0160 u

mass of ^4_2He = 4.0026 u

mass of ^1_0n = 1.0087 u

Calculate ΔE the energy released, in joules, when one nucleus of helium is created.

(Total 4 marks)
Edexcel GCE Physics Module Test PH2, January 1998

21 The diagram shows a beam of protons of mass m and charge q travelling at speed v and a beam of alpha particles travelling at the same speed v. They both enter a region where a uniform magnetic field B acts perpendicular to their direction of travel.

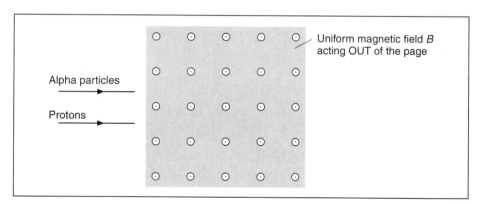

Write down an expression for the force required to accelerate the protons in a circular path of radius r_p and an expression for the magnetic force acting on the protons. Use these two expressions to show that $q/m = v/Br_p$ [3]

Add to the diagram the paths of the protons and the alpha particles after they enter the magnetic field.   [2]

What is the value $r_\alpha : r_p$ where r_α is the radius of the path of the alpha particles?   [2]

Add to the diagram the path of a beam of neutrons and that of a beam of electrons travelling with the same speed as the protons when they enter the magnetic field. **[3]**

(Total 10 marks)

Edexcel GCE Physics Module Test PH4, January 1999

22 List the main components of a circular accelerator. **[4]**

In 1995 scientists at CERN noticed slight periodic variations in the energy of the particle beams. These variations occurred at the same frequency as the tides. It was decided that this was due to gravitational forces altering the shape of the Earth and hence the dimensions of the circular accelerator. Suggest why small changes in the circumference of the accelerator affect the energy of the particle beams. **[2]**

(Total 6 marks)

Edexcel GCE Physics Module Test PH4, January 2000

Things you need to know

Chapter 1 Gravitational fields
mass: amount of matter in a body

weight: gravitational force from Earth acting on a body

field (or **force field**): region in which a force acts

gravitational field: region where gravitational forces are exerted on a mass

gravitational field strength: force exerted by a gravitational field on each kilogram

Chapter 2 Newton's law of gravitation
Newton's law of gravitation: gravitational force between two bodies is directly proportional to the product of their masses and inversely proportional to the square of their separation ($F = GmM/r^2$)

universal gravitational constant: the constant that applies in all situations of gravitational attraction

Chapter 3 Satellites
geostationary (or **geosynchronous**) **satellite:** orbits above the equator with a period of 24 h and thus maintains the same position above the Earth's surface

Chapter 4 Gravitational field lines
inverse square law: when a quantity decreases in proportion to the square of the increasing distance

Chapter 5 Coulomb's law
electronic charge: the size of the charge on either a proton (positive) or an electron (negative); 1.6×10^{-19} C

Coulomb's law: force between two charges is directly proportional to the product of their charges and inversely proportional to the square of their separation ($F = kqQ/r^2$)

Chapter 6 Radial electric fields
electric field: region where there are electric forces on charges

electric field strength: force exerted by an electric field on each coulomb

Chapter 7 Uniform electric fields
equipotential: line joining points of equal potential energy

potential gradient: rate at which potential difference changes with distance

Chapter 8 Comparing gravitational and electric fields
escape speed: vertical speed required to totally leave the gravitational influence of the Earth

Chapter 9 Charging a capacitor
capacitor: electrical device used to store energy by means of a displacement of charge within it

Chapter 10 Storing charge
capacitance: the ability of a component to 'store' charge; farad = coulomb per volt

Chapter 11 Exponential decay – capacitors
time constant: time for capacitor's charge and potential difference to decrease to 1/e of its original value

Chapter 12 Exponential decay – radioactivity
exponential decay: occurs when the rate of change of a quantity depends on how much of that quantity is present, e.g. current on amount of charge and activity on the number of undecayed nuclei

THINGS YOU NEED TO KNOW

Chapter 15 Magnets

magnetic field: region where magnetic forces are experienced

neutral point: position within overlapping magnetic fields where the resultant field is zero

Chapter 16 Magnetic effects of currents

corkscrew rule: the magnetic field is clockwise around a wire carrying a current away from you

solenoid: a cylindrical current-carrying coil of wire with a large number of turns

Chapter 17 Fleming's left-hand rule

Fleming's left-hand rule: using your left hand – first finger shows direction of magnetic field, second finger current and thumb the thrust (or force) on the conductor

Chapter 18 Magnetic field strength

magnetic flux density: a measure of the strength of a magnetic field

tesla: unit of magnetic flux density, where 1 T produces a force of 1 N on each metre length of wire carrying a current of 1 A perpendicular to the magnetic field

Hall probe: device for comparing magnetic flux densities of steady fields

Chapter 19 Laws of electromagnetic induction

magnetic flux: total amount of magnetism through an area, equal to the product of the magnetic flux density and the area

magnetic flux linkage: product of magnetic flux and the number of turns

Faraday's law: magnitude of induced e.m.f. in a circuit is directly proportional to the rate of change of magnetic flux linkage through that circuit

Lenz's law: any current driven by an induced e.m.f. opposes the change causing it

Chapter 20 Applications of electromagnetic induction

direct current: a current that flows in one direction only

alternating current: a current whose direction continually reverses

Chapter 21 Electron beams

electron gun: device for producing and projecting a beam of electrons

thermionic emission: the freeing of electrons from a metal due to increased thermal vibration of the lattice

linac: linear accelerator – contains a series of plates used to accelerated charged particles progressively along a straight path

Chapter 22 Magnetic force on a charged particle

cyclotron: device for progressively increasing the speed of charged particles in a circular path

Chapter 24 Energy has mass

fission: process of splitting up large nuclei into much lighter nuclei to produce large amounts of energy

fusion: process of joining together light nuclei to produce slightly heavier nuclei and large amounts of energy

Equations to learn

Gravitational force between two masses m and M

$F = GmM/r^2$

Electric force between two charges q and Q

$F = kqQ/r^2$

where for free space (or air) $k = 1/(4\pi\varepsilon_0)$

Capacitance $\qquad C = Q/V$

Index

Page references in **bold** refer to definitions/explanations in the 'Things you need to know' section.

INDEX

neutral points, magnetic fields 35, **75**

Newton, Isaac 4, 5, 7

Newton's law of gravitation 4–5, 8, 18, **74**

Newton's second law 3

north poles, bar magnets 34, 35

nuclear fission 55

nuclear fusion 55
  and Sun 55

ohms 25

orbital radii, planets 7

particle detectors 52–3

particle showers 52

particles
  acceleration 49, 51
  alpha 52, 53
  beta 53

periods, planets 7

permeability of free space 41

planets
  orbital radii 7
  periods 7

plutonium 55

potential, electrical 17

potential energy, electron beams 48

potential gradient 15, **74**

proton beams 52

protons 10
  acceleration 52

radial electric fields 12–13, 18

radial fields, potential gradient 16

radiation
  alpha 53
  beta 53
  gamma 53

radioactive decay 24
  compared to capacitor discharge 27
  equations 26–7

repulsive electrostatic forces 11, 18

resistance and capacitors 23, 25

resistors, compared to capacitors 31

satellites
  artificial 7
  geostationary (geosynchronous) 7
  and gravitational field strength 6
  near-Earth 7
  and telecommunications 7

shields 18

solenoids 37, 41, **75**
  and bar magnets 37

south poles, bar magnets 34, 35

spark chambers 52–3

speed of light 49

springs, energy storage 28

Stanford linear accelerator 49

Sun
  gravitational fields 2
  and nuclear fusion 55

synchrotrons 52

telecommunications and satellites 7

teslas 40, 42, **75**

thermionic emission 48, **75**

time constant, exponential decay 24–5, **74**

tracks, particles 52, 53

transformers 46

unified mass unit 54

uniform electric fields 18
  equipotentials 14, 15
  potential gradient 15

uniform gravitational fields 9, 18
  work done 9

universal gravitational constant 4, 8, **74**
  measurement 5

uranium 55

Van de Graaff generator 49

voltage, alternating 46

voltage/e.m.f., induced 42, 43, 44, 46

webers 42

weight **74**
  and gravitational forces 2, 4
  measurement 2

work done
  electric fields 14
  energy storage 28–9, 31
  uniform gravitational fields 9